MACHINISTS
LIBRARY
Volume II

Basic Machine Shop

by Rex Miller

THEODORE AUDEL & CO.
a division of
THE BOBBS-MERRILL CO., INC.
Indianapolis/New York

FOURTH EDITION
FIRST PRINTING

Copyright © 1965, 1970, and 1978 by
Howard W. Sams & Co., Inc.
Copyright © 1983 by The Bobbs-Merrill Co., Inc.

Published by The Bobbs-Merrill Company, Inc.
Indianapolis/New York
Manufactured in the United States of America

Library of Congress Cataloging in Publication Data

Miller, Rex, 1929-
    Machinists library.

    Revision of the 3rd ed. of three volumes previously published separately in
1978 under titles: Machinist library, basic machine shop, Machinists library:
machine shop, and Machinists library: toolmakers handy book.
    Includes indexes.
    Contents: v. 1. Basic machine shop—v. 2. Machine shop,—V. 3. Toolmakers
handy book.
    1. Machine-shop practice   I. Title.
TJ1160.M566   1984   621'.9'02   83-6383
ISBN 0-672-23381-9 (v.1)
ISBN 0-672-23382-7 (v.2)
ISBN 0-672-23383-5 (v.3)
ISBN 0-672-23380-0 (set)

Foreword

The purpose of this book is to provide a better understanding of fundamental principles of machine shop practices and operations for those persons desiring to become machinists. The beginning student or machine operator cannot make adequate progress in his trade until he possesses a knowledge of the basic principles involved in the proper use of various tools and machines commonly found in the machine shop.

One of the chief objectives has been to make the book clear and understandable to both students and workers; therefore, illustrations and photographs have been used generously to present the how-to-do-it phase of many of the machine shop. The material presented should be helpful to the machine shop instructor as well as to the individual students or workers who desire to improve themselves in their trade.

The proper use of machines, as well as the safety rules for using them, has been stressed throughout the book. Basic principles of setting the cutting tools and cutters are dealt with throroughly, and recommended methods of mounting the work in the machines are profusely illustrated. The role of numerically controlled machines is covered in detail with emphasis upon the various types of machine shop operations that can be performed by them.

This book is presented at a time when there is a definite trend toward expanded opportunities for vocational training of young adults. An individual who is ambitious to perfect himself in the machinist trade will find the material presented in an easy-to-understand manner whether he is a young man studying alone in his spare time or an apprentice working under close supervision on the job.

REX MILLER

Acknowledgments

A number of people and companies have been responsible for the extensive illustrations in this book. I would like to thank them for their contributions nad personal efforts on behalf of this publication. Each contributor has been identified with the drawing or photo shown. Thanks for your time and efforts.

Contents

CHAPTER 23

CHAPTER 24

CHAPTER 25

CHAPTER 1

Power Hacksaws, Power Bandsaws, and Circular Saws

Power hacksaws, power bandsaws, and circular saws are very important to machine shop operations. A large number of power hacksaws and power bandsaws are in use in the metalworking industry.

POWER HACKSAWS

A power hacksaw is an essential machine in most machine shop operations. For many years a hand-operated hacksaw was the only means for sawing off metal. Power-driven machines for driving metal-cutting saw blades have been developed to make the task easier. The power hacksaw can do the work much more rapidly and accurately. The machinist should be familiar with these machines, the blades used on the machines, and the operations performed on them. One type of power hacksaw commonly found in machine shops is shown in Fig. 1-1.

Fig. 1-1. Keller automatic lift power hacksaw with coolant system.

Basic Construction

Power hacksaws are designed to make the sawing of metal a mechanical operation. The stock is usually held in a vise mounted on the base of the machine. An electric motor is used to supply power for the machine.

Drive Mechanism—The drive shaft is connected by a V-belt and gears to the electric motor mounted on the machine. The drive mechanism is shielded by guards for safety in operation.

Frame—A U-shaped frame is used on the smaller power hacksaws to support the two ends of the saw blade, which is under tension. The heavier machines use a four-sided frame and a thin backing plate for the blade of the saw.

Worktable and Vise—Most worktables are equipped with a vise that can be mounted either straight or angular to the blade. The worktable is usually mounted on a ruggedly constructed base. Many worktables are provided with T-slots for the purpose of supporting special clamping devices.

Special Features—Nearly all power hacksaws raise the blade on the return stroke. This feature prevents dulling of the blade by dragging it over the work as the blade is returned to the starting position.

Another important feature is a blade safety switch that automatically stops the machine if the blade should break during operation of the saw. The safety switch prevents any damage that could result if the machine continued operation with a broken blade.

Coolant System—Some power hacksaws are equipped with a coolant system that delivers a coolant to the hacksaw blade. The coolant passes from a receiving tank to a pump and then to the work. The machine is equipped with a trough to catch the coolant, which may be screened to remove any chips of metal (see Fig. 1-1).

Saw Capacity—Small power hacksaws can be used on square or round stock ranging from ⅛ inch to 3 inches. The larger machines have a capacity ranging to 12 inches, square or round, or even larger.

The capacity of a machine for angular cuts is different from its capacity for straight cuts. The cutting surface is longer for angular cutting. Thus, the saw must be equipped, not only with a swivel vise, but it also must have a long enough stroke to make the angular cut.

Blades

High-speed tungsten steel and high-speed molybdenum steel are the most commonly used materials in power saw blades. If only the teeth are hardened, the blades are called flexible blades.

Power hacksaw blades are ordered by specifying length and width, thickness, and teeth per inch. For instance, they are available in 12″ × 1″ length and width at 0.050-inch thickness. The teeth per inch (TPI) would be either 10 or 14. The 14″ × 1″ × 0.050″ blade would be available in only 14 teeth per inch. Blades also come in 17″ × 1″ with a thickness of 0.050 inch and either 10 or 14 teeth per inch. Blades with a thickness of 62 thousandth (0.062 inch) are usually 1¼ inches wide.

High-speed, shatter-proof blades are designed to meet safety and performance requirements. The high-speed molybdenum blades are longer wearing and give the best results for general use. They, too, come in 12-inch, 14-inch, and 17-inch lengths with a 1-inch or 1¼-inch width. These blades are made in both 10-and 14-teeth-per-inch sizes. Thickness of the metal being cut determines the number of teeth per inch choen to do the job. There should be no fewer than 2 or 3 teeth touching the metal being cut.

Fig. 1-2 shows the power hacksaw blade end with a hole for mounting in the machine and the pointed nature of the teeth.

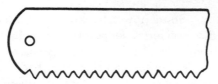

Fig. 1-2. Power hacksaw blade.

Operation

Straight cuts are made easily on power hacksaws. The visae is stationary, and the cut is made at a right angle to the sides of the stock.

Most machines are equipped with an adjustable vise. Angular cuts at any desired angle up to 45 degrees can be made by swiveling the vise.

To operate the power saw properly, the work should be fastened securely in the vise so that the blade will saw in the proper place. The blade will break if the work loosens in the vise.

The saw blade should be lowered onto the work carefully to start the cut. On some machines this is done by hand, but it can be done automatically on some saws. In either method, the points of the teeth will be broken or damaged if the blade is permitted to strike the work suddenly.

The machine should be watched carefully to make certain that the saw blade lifts about ⅛ inch on the return stroke. If the blade fails to lift, adjustments should be made immediately, as the blade will be damaged if operation is continued.

Another precaution is to be certain, when making angular cuts with the work turned at an angle in the vise, that the saw blade can make both the backward and the forward strokes without the saw frame making contact with either the work or the vise. Serious damage to the machine can result from failure to observe this precaution.

When a saw blade is replaced, or a new blade is started in an old cut, it should be remembered that the set of the new blade is wider. The new blade will stick in the old cut unless the work is rotated in the vise a quarter-turn. If the work cannot be rotated, the new blade should be guided into the old cut.

The cutting speed of a power hacksaw, of course, varies with the material being cut. Suggested cutting speeds are as follows: For mild steel, 130; for tool steel (annealed), 90; and for tool steel (unannealed), 60. For example, on a machine with a 6-inch stroke, the revolutions per minute of the driving crank should be 130.

All steels should be cut with a cutting compound. Bronze should be cut with a suitable compound at the same speed as mild steel. The saw blade will heat rapidly if an attempt is made to cut brass without a cutting compound adapted to brass. Brass may be cut at the same speed as steel if a suitable compound is used.

POWER BANDSAWS

In the past few years the power bandsaw has become very important in machine shop operations. In some instances it is used in production operations prior to final machining operations.

Basic Construction

Power bandsaws are also designed to make the sawing of metal a mechanical operation, see Fig. 1-3. The stock can be held in a vise mounted on the machine or it can be supported by the operator's hand. Electric motors are used to supply the power for the bandsaws.

Drive Mechanism—Wheels on the power bandsaw can be adjusted to apply tension to the bandsaw blade, which is a flexible, thin, narrow ribbon of steel. One of these wheels is powered by the electric motor mounted on the machine. These wheels are enclosed by guards for safety in operation.

Frame—Many variations of power bandsaws are available. However, bandsaws can be grouped into three classifications: horizontal machines for cut-off sawing; vertical machines for straight and profile sawing at conventional speeds; and vertical machines for nonferrous cutting and friction cutting.

Worktable and Vise—Power bandsaws have either a worktable or vise to hold the metal that is to be cut. The worktable is usually

Fig. 1-3. A power bandsaw.

part of the bandsaw. The vise, if there is one, is usually designed to fit the worktable. Many worktables also have T-slots for the purpose of supporting special clamping devices to hold the work to be cut.

Special Features—Some power bandsaws have automatic tensioning devices so that the proper amount of tension is applied to the bandsaw at all times. This type of device reduces excessive wear and damage to the blade.

Coolant System—Coolant systems are common on power bandsaws designed for high-speed production work. These systems deliver coolant to the bandsaw blade and the work. After the coolant has been used, it is recycled through a screen and filter to remove any chips.

Capacity of Power Bandsaws—The maximum capacity of power bandsaws varies according to the size of the bandsaw.

Some of the larger machines can accommodate work that is 18″ × 18″ and larger. Bandsaws also have the capacity to do a wide variety of operations that include cut-off, straight sawing, and contour, or profile, sawing. Rough shaping and semifinishing can be accomplished on almost all types of ferrous and nonferrous materials.

Blades—The development of the heavy-duty bandsaw machine tool has been designed and built to operate with high-speed blades at maximum efficiency. Use of the bandsaw offers several important advantages over the hacksaw. These advantages include a narrower kerf, which results in greatly reduced losses; fast, efficient cutting with low per-cut costs; and consistently smooth, accurate cutting.

The type of blade used on the power bandsaw is very important. Blades are available with many different types of teeth, tooth set, and pitch.

The three types of teeth are: the regular tooth, skip tooth, and hook tooth. See Fig.1-4. The regular tooth blade provides from 3 to 32 teeth per square inch. Teeth may be raker set or wavy set. Regular tooth blades are preferred for all ferrous metals and for general purpose cutting.

Skip-tooth blades feature widely spaced teeth, usually from 2 to 6 teeth per inch, to provide the added chip clearance needed for cutting softer materials. Teeth of the skip-tooth blade are characterized by straight 90 degree faces and by the sharp angles at the junction of the tooth and gullet. Since these sharp angles and flat

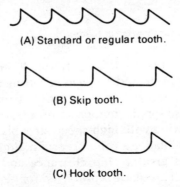

(A) Standard or regular tooth.

(B) Skip tooth.

(C) Hook tooth.

Fig. 1-4. Three tooth styles for blades.

17

surfaces tend to break up chips, this type of tooth is preferred for the very soft nonferrous metals that would otherwise tend to clog and gum the blade. This blade is also widely used in wood cutting.

Hook-tooth blades provide the same wide tooth spacing as the skip-tooth blade. The teeth themselves have a 10 degree undercut face. Gullets are deeper with blended radii between the teeth and the gullets. Since the undercut face helps the teeth dig in and take a good cut, the blended radii tend to curl the chips. This type of tooth is preferred for the harder nonferrous alloys and many plastic operations.

There are three basic types of tooth set on metal-cutting band-saw blades. Each tooth set is designed for a specific type of cutting application. These three basic types of tooth sets are: the raker set, the wavy set, and the alternate set. See Fig. 1-5.

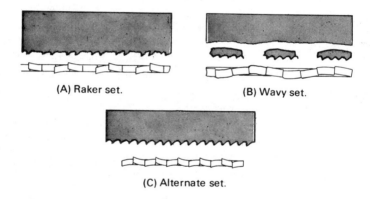

(A) Raker set. (B) Wavy set.

(C) Alternate set.

Fig. 1-5. Types of tooth sets for blades.

The raker set—one tooth set left, one tooth set right, and one tooth not set—is preferred for all long cutting runs where the type of material and the size and shape of the work remain relatively constant. It is also for all contour, or profile, cutting applications, and for bandsawing with high-speed steel blades.

The wavy set blade, a rolled set with alternate left and right waves, provides greater chip clearance and a stronger, nearly rip-proof tooth. This blade is preferred for general-purpose work on conventional horizontal machines.

The alternate set has every tooth set—one to the left, one to the right—throughout the blade. Taken originally from the carpenter handsaw, it was supplied in bandsaws for brass foundry applications.

Operation

When operating a bandsaw, there are some general principles that should be followed so that you will be able to select the blade with the proper pitch. See Fig.1-6 for any cutting application.

1. Small and thin-wall sections of metal require fine teeth.
2. Large metal sections require the use of blades with coarse teeth so that adequate chip clearance is provided.
3. Two teeth should be engaged in the metal to be cut at all times.
4. Soft, easily machined metals require slightly coarser teeth to provide chip clearance. Hard metals of low machinability require finer teeth so that there are more cutting edges per inch.
5. Stainless steel should be sawed with a coarse tooth sawblade for best results.

(A) Correct and incorrect tooth pitches.

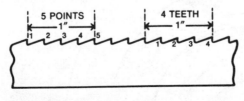

(B) How determined.

Fig. 1-6. Tooth pitches.

Select the widest blade that your machine will accommodate when straight, accurate cutting is required. Blade widths for contour sawing depend upon the specific application.

Profile, or contour, sawing is an accurate, fast, and efficient method of producing complex contours in almost any machinable metal. With the proper blade selection, radii as small as $\frac{1}{16}$ inch can be cut. Either internal or external contours can be sawed. If an internal contour has to be sawed, it is first essary to drill a hole within the contour area to accommodate the saw blade. The bandsaw blade has to be cut to length, threaded through the drilled hole, then rewelded. It is for this reason that bandsaws designed and built for contour sawing have a built-in butt welder. Blades for profile sawing are always raker set because this type of tooth provides the necessary side clearance. These blades are usually narrower than those chosen for straight cutting; the extra width for any specific job depends upon the smallest radius to be cut. Table 1-1 will aid you in selecting the correct blade for any application.

Table 1-1. Bandwidth Selection for Profile or Contour Sawing

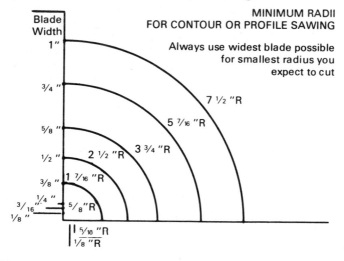

MINIMUM RADII
FOR CONTOUR OR PROFILE SAWING

Always use widest blade possible
for smallest radius you
expect to cut

Profile sawing requires cutting and rewelding the blade. Because of this, it is necessary to make a good weld. This can be accomplished very easily if certain steps are followed. The steps are:

1. Square both ends of the blade accurately.
2. Align both ends of the blade so that they join evenly across the entire width of the blade.
3. Select the correct dial setting for the blade width.
4. Select the correct annealing period.
5. Dress the blade to remove the welding flash.
6. Anneal for stress relieving.

With the exception of cast iron, coolant should be used for all bandsawing operations. Cast iron is always cut dry. Almost all of the commercially available soluble oils or light cutting oils will give good results when cutting ferrous metals. When aluminum is cut, paraffin or beeswax are commonly used lubricants. See Table 1-2.

Table 1-2 is based on four types of cutting fluids:

A. *Cutting Oil Mineral Base with Fatty Oils Added.* Has Sulphur for antiweld characteristics and Chlorine for film strength.
B. *Cutting Oil with Light Viscosity Fatty Oils.* For cold weather operations mix with (A). When cutting copper alloys, additives and Chlorine should be added.
C. *Synthetic Water Soluble Cutting Agent.*
D. *Soluble Oil Cutting Agent* with fatty oils and sulphurized for extreme pressure with antiweldment properties. Water soluble for greater heat removal.

The three types of feeds used on power bandsaws are: hand, hydraulic, and mechanical. The number of factors involved in the efficient and troublefree operation of a power bandsaw make it difficult to provide specific information on the proper speed or pressure required. See Table 1-3. When a cut is started, the blade should not be forced. If the work is forced into the blade, it will result in a shorter blade life and defective work. The following principles should be used as a guide in selecting the proper feed pressure:

21

Table 1-2. Recommended Cutting Fluids

OPERATION	MATERIALS TO BE CUT—CUTTING FLUIDS RECOMMENDED					
	Free Cutting Steels Carbon Steels Cast Steels Malleable Iron	Alloy Steels Nickel Steel Nickel Chromium Steel	Stainless Steels (Free Cutting) Ingot Iron Wrought Iron	Stainless (Austenitic) Manganese Steels Hi-Temp Alloys	Aluminum Alloys Leaded Brass Magnesium Alloys Phosphor Bronze Zinc	Brass Gun Metal Nickel Monel Inconel
Production Sawing Using MARVEL High-Speed Steel, High-Speed-Edge, and Intermediate Alloy Steel Band Blades	FLOOD (B) (C) Mix 1 to 5 with water (D) Mix 1 to 5 with water	FLOOD (A) (C) Mix 1 to 5 with water (D) Mix 1 to 5 with water	FLOOD (A) (D) Mix 1 to 5 with water (C) Mix 1 to 5 with water	FLOOD (A) (C) Mix 1 to 5 with water (D) Mix 1 to 5 with water	FLOOD (B)	FLOOD (A) (C) Mix 1 to 5 with water (D) Mix 1 to 5 with water
General Sawing Using MARVEL Hard Back & Flexible Back Hardened Edge Band Blades	DRIP (Blade just wet) (B) (D)	DRIP (Blade just wet) (B) (D)	DRIP (Blade just wet) (B) (D)	DRIP (Blade just wet) (A) (B)	DRIP (Blade just wet) (B) Paraffin Dry	DRIP (Blade just wet) (A) (B)
Contour Sawing Using MARVEL High-Speed-Edge or Hard Back & Flexible Back Carbon Blades	SPRAY OR DRIP (C) (D) Mix 1 to 10 with water	SPRAY OR DRIP (C) (D) Mix 1 to 10 with water	SPRAY OR DRIP (C) (D) Mix 1 to 10 with water (B)	FLOOD OR DFIP (D) Mix 1 to 10 with water (C) Mix 1 to 1C with water (A)	FLOOD (B) Paraffin Dry	(B) (A)

Courtesy Armstrong-Blum Mfg. Co.

1. Use moderate feed pressure when straight, accurate cutting is desired.
2. Avoid heavy feed pressures as they will cause the power bandsaw to chatter and vibrate.
3. Study the chips produced. Fine powdery chips indicate the feed is too light. The teeth on the blade are rubbing over the surface of the work instead of cutting the work. Discolored, blued, or straw-colored chips indicate that there is too much feed pressure. This excess pressure can cause teeth to chip and break. In addition, the blade will wear prematurely due to overheating. A free-cut curl indicates ideal feed pressure with the fastest time and the longest blade life.

One of the most important factors in successful bandsawing is a proper cutting speed. If the machine is operated at too fast a speed

Table 1-3. Common Band Sawing Problems and Their Connections

PROBLEM	CAUSE	CORRECTION
BLADE DEVELOPS CAMBER	Roller guides not correctly adjusted.	Readjust.
	Feeding pressure is too heavy.	Use lighter feed.
	Saw guides too far apart.	Adjust closer to work.
	Blade riding against flange on metal wheel.	Tilt wheel for proper blade position.
BLADE DEVELOPS TWIST	Saw is binding in cut.	Decrease feeding pressure.
	Side inserts or rollers of saw guides too close to saw.	Readjust.
	Wrong width of blade for radius being cut.	Check for correct blade width.
SAW DULLS PREMATURELY	Saw speed too great; teeth sliding over work instead of cutting.	Reduce saw speed.
	Improper coolant or coolant improperly directed.	Check type and mixture of coolant. Apply at the point of cut, saturating teeth evenly.
	Saw idling through cut.	Keep teeth engaged; use positive feeding pressure.
	Feed too light; teeth sliding over work.	Use a feed heavy enough to generate a full curled chip.
SAW LOSES SET PREMATURELY	Saw is too wide for radius being cut.	Check for the proper width.
	Saw speed is too fast.	Reduce speed.
	Saw rubbing against vise or running deep in guides.	Check blade along complete travel.

Table 1-3. Common Band Sawing Problems and Their Connections (Cont'd.)

SAW VIBRATES IN CUT	Wrong speed for the material and work thickness.	Check for the proper width.
	Insufficient blade tension.	Increase tension.
	Pitch too coarse.	Select a finer pitch.
	Excessive feeding pressure.	Reduce pressure.
SAW TEETH RIP OUT	Pitch too coarse.	Use a finer pitch on thin work sections.
	Work not tightly clamped or held.	Re-clamp or hold more firmly to prevent vibration.
	Sawing dry.	Use coolant when possible.
	Gullets loading.	Use a coarser pitch and/or a higher viscosity lubricant—or brush to remove chips.
	Excessive feed pressures.	Reduce.
SAW BREAKS PREMATURELY	Blade too thick for the diameter of wheels.	Check recommendations for your machine.
	Cracking at the weld.	Weld improperly made; try a longer annealing period.
	Pitch is too coarse.	Use finer pitch.
	Excessive feeding pressure or blade tension.	Reduce.
	Guides too tight.	Readjust.

Courtesy Henry G. Thompson Co., Subsidiary of Vermont American Corporation

for the material being cut, the teeth are not allowed sufficient time to dig into the material. As a result, they merely rub over the surface of the work, creating friction that rapidly dulls the cutting edge and wears out the blade. Table 1-4 gives average speeds and cutting rates for cutting a wide variety of commonly used ferrous and nonferrous metals and nonmetallics. These recommendations will ensure optimum performance under most conditions.

Friction cutting differs from all other types of metal sawing methods. It is not actually a cutting operation, but a burning process similar to torch cutting. It is much faster than conventional sawing methods. Very hard materials, which could not normally be sawed, are cut rapidly and easily. Bulky and irregular shapes can be cut handily because there is little blade drag.

Basically, the friction cutting operation involves the use of a very fast moving blade that travels at speeds between 6000 and

18,000 feet per minute. See Table 1-5. At these speeds, terrific heat is built up in the workpiece at its point of contact with the blade, and burning begins. Teeth are not needed to generate burning heat, but they generate increased cutting speed and efficiency by carrying additional oxygen into the work area, thus creating

Table 1-4. Average Speeds and Cutting Rates

FLEXIBLE AND HARD BACK CARBON—RAKER TOOTH								
MATERIAL	SIZE OF MATERIAL							
	1/4"		1/2"		2"		4"	
	Speed	Teeth	Speed	Teeth	Speed	Teeth	Speed	Teeth
STEELS								
Armor Plate	150	18	125	12	75	8	50	6
Angle Iron	175	24	150	14				
Carbon Steels	250	24	200	14	150	10	100	6
Chromium Steels	150	24	125	14	100	10	50	6
Cold-rolled steel	250	12	200	10	150	8	125	6
Drill rod	100	14	100	14				
Graphite steels	175	18	150	14	100	10	75	6
High speed steels	150	24	100	14	75	10	50	8
Machinery steels	250	18	200	14	150	10	125	6
Molybdenum steels	150	18	125	14	75	10	50	8
Nickel steels	150	18	125	14	75	10	50	8
Silicon manganese	175	18	150	14	75	10	50	6
Stainless steels	100	24	100	14	50	10	50	6
Structural steels	175	24	150	14				
Tungsten steels	175	18	150	14	75	8	50	6
FOUNDRY METALS								
Brass—hard	500	18	400	14	300	10	200	6
Brass—soft	1500	18	1000	14	750	10	300	6
Bronze—aluminum	500	18	400	14	225	10	100	6
Bronze—manganese	300	18	250	14	200	10	150	6
Bronze—naval	300	18	275	14	225	10	150	6
Bronze—phosphorous	500	18	400	14	200	10	150	6
Cast iron—gray	200	18	175	14	100	8	75	6
Cast iron—malleable	200	18	175	14	150	10	125	6
Cast steel	225	18	200	14	100	8	75	6
Copper—beryllium	400	18	350	12	250	10	150	6
Copper—drawn	1100	18	700	10	350	6	200	6
Gunnite	300	24	200	18	150	10	100	6
Meehanite	150	18	100	14	75	8	50	6
Monel	200	18	150	14	75	10	50	6
Nickel—cold-rolled	200	14	150	10	75	8	50	6
Nickel silver	250	18	250	14	175	10	125	6
Silver	250	24	250	18	175	10	150	6

Table 1-4. Average Speeds and Cutting Rate (Cont'd.)

MATERIAL	FLEXIBLE AND HARD BACK CARBON—HOOK TOOTH							
	SIZE OF MATERIAL							
	½"		2"		5"		10"	
	Speed	Teeth	Speed	Teeth	Speed	Teeth	Speed	Teeth
NON-FERROUS METALS								
Aluminum—soft	4000	4	3600	3	3200	3	3000	2
Aluminum—medium	3000	6	2000	4	1500	3	1200	2
Aluminum—hard	600	6	500	4	400	3	300	3
Babbit	4000	6	3500	4	3000	3	2500	3
Beryllium copper	2000	6	1800	4	1400	3	1000	3
Brass-casting	3200	6	3000	4	2500	3	1800	3
Brass-commercial	3700	4	3500	3	3000	3	2500	2
Brass—naval	3500	4	3000	3	2500	3	2000	2
Brass—yellow	3200	4	3000	3	2500	3	1800	2
Cadmium—Kirsite	3200	6	3000	4	2500	3	1700	3
Copper	3000	4	3000	3	2500	3	2000	2
Lead	4000	6	3500	4	2700	3	2000	3
Magnesium	4000	6	3000	4	2500	3	2000	2
Silicon bronze	1000	6	1000	4	600	3	300	3
Titanium #150A	60	6	50	6	50	4	50	3
STEELS								
Alloy steels	80	6	70	6	60	4	50	3
Low carbon steel	170	6	150	4	125	3	100	2
Medium carbon steel	120	6	110	4	100	3	90	2
High carbon steel	90	6	80	6	70	4	65	3

Table 1-5. Recommended Friction Cutting Speeds for Common Metals

DESCRIPTION	Average Recommended Cutting Speeds in Feet per Minute
Armor Plate	7,000–13,000
Carbon Steel	6,000–12,000
Cast Steel	7,000–13,000
Chromium Steel	8,000–15,000
Chromium Vanadium Steel	8,000–15,000
Free Machining Steel	6,000–12,000
Gray Cast Iron	7,000–13,000
Malleable Cast Iron	7,000–13,000
Manganese Steel	6,000–12,000
Moly Steel	8,000–15,000
Molybdenum	8,000–15,000
Nickel Chromium Steel	8,000–15,000
Nickel Steel	8,000–15,000
Silicon Steel	8,000–15,000
Stainless Steel	7,000–13,000
Tungsten Steel	8,000–15,000

greater oxidation. Set teeth are also important. They give the operator the control that he must have to cut a straight line or follow a specific curve.

There is a definite advantage to using a special blade for this type of operation. The friction cutting blade is made from special steel. The blade is specially heat treated to withstand the fast speeds and severe flexings that are encountered in friction sawing.

Many difficult manufacturing problems have been solved with friction sawing because this process has distinct advantages. Before deciding whether to use friction sawing, consideration should be given to the specific job requirements. There are important limitations involved with friction sawing. Some of these limitations are:

1. Friction cutting leaves a very heavy burr on the underside of the material being cut.
2. Thicknesses over ⅝ inch are extremely difficult to cut.
3. Heat from the friction-sawing process may have a tempering effect on the edges of some materials.

CIRCULAR SAWS

Circular saws have some advantages. They produce a burr-free mill finish when slow-speed cold sawing is used. This eliminates secondary operations on tubing, channels, angles, and solid stock of most steels and other ferrous materials, as well as most nonferrous metals. The rigidity of the blade produces cuts of extreme accuracy and close tolerances. The cutting operation is safe, clean, and quiet because of the slow r/min.

Fig. 1-7 shows a manually operated vise model saw and a semiautomatic type with an air-operated vise. Special vise insets for holding thin-walled pipe or tubing to prevent distortion are shown in Fig. 1-8A. The other holding arrangement is shown in Fig. 1-8B. This shows square material being held for optimum cutting angle. Other special shapes can be made to fit in the vises, or they can be purchased from the saw manufacturer.

The semiautomatic machines have heavy-duty feed mechanisms to feed the metal to the saw. The saw can do straight cutting,

(A) manual

(B) semiautomatic

Courtesy Startrite, Inc.

Fig. 1-7. Slow-speed circular saws: (A) manual; (B) semiautomatic.

miter cutting, slot, or longitudinal, cutting, or any number of other arrangements that will fit within the limits of the machine.

Nonferrous Saw

A machine specifically suited for aluminum or nonferrous and hard plastics is shown in Fig. 1-9. The machine cuts from the bottom up. The cutting blade speed is 9840 surface feet per minute (sfpm). The workpiece is held from the top by a quick-clamping vise. Maximum saw diameter is 12 inches. A built-in spray mist coolant system provides an effective blade lubrication. Mitering is possible in either 45 degrees or 60 degrees and in both directions.

Blade—The circular saw blade is the most important part of the saw. It is chosen to do a specific job. The blade is important for the quality of the cut. High-speed blades are coated with a special

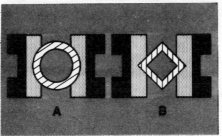

Fig. 1-8. Special vise inserts.

nitrate treatment to retain strength and hardness, as well as to facilitate coolant penetration. Carbide-tipped blades are made of a special alloy with carbide-tipped teeth for high-speed cutting of aluminum and other light nonferrous materials.

Table 1-6 shows the approximate number of teeth for the blade diameter. This table indicates the actual number of teeth engaged with the material being cut. No fewer than 2 to 3 teeth should engage the work at all times. No more than 7 teeth should be engaging the work at any one time.

Table 1-6. Approximate Number of TPI

Number of Teeth on Blade	Blade Diameter					
	6"	8"	10"	11"	12"	14"
80	4	3	2½	2	2	1½
100	5	4	3	3	2½	2
120	6	5	4	3½	3	2½
140	7	5½	4½	4	3½	3
160	8	6	5	4½	4	3½
180	9	7	6	5	5	4
200	11	8	6½	6	5½	4½
220	12	9	7	6½	6	5
240	13	10	7½	7	6½	5½
260	...	...	8	7½	6	...
280	...	...	9	8	7	...
300	...	...	10	9	8	...

Table 1-7 shows the blade pitch and the number of teeth per inch; Table 1-8 indicates the blade speed in feet per minute for the different spindle speeds and blade diameters.

Fig. 1-9. High-speed aluminum and nonferrous saw.

Table 1-7. Pitch Selection

Pitch Designation	2	3	4	5	6	8	10	12
Approx. Teeth Per Inch	14	10	6	5	4	3	$2\frac{1}{2}$	2

Table 1-8. Blade Speeds-FPM

Spindle Speed	Blade Diameter					
RPM	8″	10″	11″	12″	13″	14″
26	54	68	75	82	89	95
52	108	136	150	164	177	190
3000	. . .	. . .	. . .	9425	. . .	. . .

Table 1-9 shows recommended sawing speeds for various materials cut by this type of saw. Table 1-10 indicates the spindle speeds available for five different models of saws. The spindle speed for the circular saws for ferrous metals are 26 to 52 r/min, or a very slow speed. The nonferrous model saw uses a speed of 3000 r/min, or a somewhat faster saw blade.

Table 1-9. Recommended Sawing Guide

MATERIAL	BLADE SPEED-FPM
Mild steel	60–165
High carbon, stainless, alloys, tool steel	40–50
Cast iron	40–125
Bronze	100–170
Brass, copper	120–170
Aluminum	150–9000

Table 1-10. Startrite Spindle Speeds

Model	Spindle Speed
CF 275	26-52 r/min
CF 300	26-52 r/min
CF 325	26-52 r/min
CF 350	23-46 r/min
CN 300 T	300 r/min

Courtesy Startrite Corp.

SUMMARY

Power hacksaws are designed to make the sawing of metal a mechanical operation. A power hacksaw can do the work much more rapidly and accurately than by using a hand tool.

Various styles of power hacksaws are manufactured to meet the many demands. Most worktables on these machines are equipped with a vise that can be mounted either straight or at an angle to the saw blade. Many worktables also provide T-shots for the purpose of supporting special clamping devices.

To operate the power hacksaw properly, the work should be fastened securely in the vise so that the blade will saw in the proper place. Many times the blade will break if the work loosens in the vise.

Nearly all power hacksaws raise the blade automatically on the return stroke. This prevents dulling the blade by dragging it over the work as the blade is returned to the starting position. Another feature of most hacksaws is the automatic safety switch, which stops the machine automatically if the blade should break during operation. The safety switch prevents any damage that could result if the machine continued operation with a broken blade.

Power bandsaws are also designed to make the sawing of metal a mechanical operation. Use of the bandsaw offers several important advantages over the hacksaw. These advantages include a narrower kerf, which results in greatly, reduced chip loses; fast, efficient cutting with lower per-cut costs; and consistently smooth, accurate cutting.

Many variations of power bandsaws are available. However, bandsaws can be grouped into three classifications: horizontal machines for cut-off sawing; vertical machines for straight and profile sawing at conventional speeds; and vertical machines for nonferrous cutting and friction cutting.

Power bandsaws have either a worktable or vise to hold the metal that is to be cut. The worktable is usually part of the bandsaw. The vise, if there is one, is usually designed to fit the worktable.

Slow-speed circular saws are used to cut ferrous metals and produce a mill-like finish free of burrs. They operate at 25 and 52 r/min with saw blades from 8 to 14 inches in diameter. The number of teeth on the saw blades varies with the diameter of the blade. Blades with 80 to 300 teeth per inch are available for use with the slow-speed saw so that it can be used to cut a number of sizes of metal accurately.

The nonferrous and aluminum saw can also be used to cut hard plastics. Mitering is possible to very accurate tolerances. Cutting speed is faster than with the ferrous-type saw. The blades are lubricated with a mist coolant system. Carbide-tipped blades are also available. When fast, accurate cuts are needed in nonferrous metals, this is the saw to use. It can be used where a fast clamping operation is needed to increase production.

REVIEW QUESTIONS

1. What material is used to make power hacksaw blades?
2. How is material held during cutting operations on a power hacksaw?
3. What is the purpose of the automatic safety switch on the power hacksaw?
4. What are two types of metal-cutting bandsaws?
5. What are the three types of teeth used on bandsaw blades?
6. What is friction cutting?

CHAPTER 2

Basic Machine Tool Operations

Machine tools are relatively new. The machine tools in use today are the products of the Industrial Revolution that started a little more than two hundred years ago. Prior to the industrial Revolution, the various items that man wore or used were made by hand tools that had changed little in their basic design for centuries. It was not until machine tools replaced hand tools that the life of abundance we have come to know and enjoy began to evolve.

Machine power was substituted for muscle power to cut, shape, and form metal products. Drilling, turning, milling, planing, grinding, and metal forming, the basic machine tool processes in use today, evolved from hand tools.

Prior to World War II, machine tools played a very important part in making the United States the most highly industrialized nation on the face of the earth. Along with this came one of the highest standards of living in the world. During World War II, the

the country's machine tool population more than doubled to meet the war production demands.

In the 1950's, the design of machine tools began to change drastically. These changes were primarily the result of the development of numerically controlled machine tools and the advancements made in the area of electronics. In the short time since then, a technological revolution has taken place in the design and development of machine tools. Advances in design and development have been so great that they have surpassed all previous advancements.

The development of this new technology did not occur by accident. This new technology is the direct result of an unprecedented upsurge in spending by the machine tool industry for research and development. The rate of technological change is accelerating more and more rapidly every year. The life span of a new machine tool is considerably shorter than it once was because the speed with which new products are constantly being developed has rapidly increased.

Once machine tools were simple and uncomplicated. This is no longer true. Machine tools are becoming more and more complex as more features are being built into their design and performance. For the machine tools we hear so much about, see Figs. 2-1 through 2-8.

Today there are unlimited varieties and combinations of machine tools. Some of these machine tools are small enough to be mounted on a table. Some machine tools are so large that they require special buildings to house them. Their cost can range from a few hundred dollars to hundreds of thousands of dollars. Some machine tools, such as presses, weigh several hundred tons. When some of the machine tools are installed, their foundations have to be extended into the ground to a depth of more than 100 feet. They can also be several stories in height. Some machine tools are incorporated into production lines hundreds of feet in length.

Whether machine tools are small or large, inexpensive or extremely expensive, they can be grouped into six major classifications. These major classifications are based upon the operations performed by the machines to shape metal. These basic operations include: drilling and boring, including reaming and tapping; turning; milling; planing, including shaping and broaching; grind-

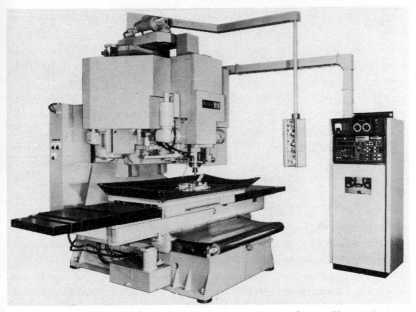

Courtesy Monarch Cortland

Fig. 2-1. A vertical spindle machining center with heavy milling capacity. This milling capacity, together with accurate boring, drilling, and tapping, expands the capability of this type of machine to accommodate an exceptionally broad range of parts. Once set up, the operation is completely automatic, with fast recycling, no lost motion and no unnecessary idle time.

ing, including honing and lapping; and metal forming, including shearing, stamping, pressing, and forging. In addition to the basic metal-shaping operations, newer metal-shaping operations have been developed during the past two decades that employ the various characteristics of chemicals, electricity, magnetism, liquids, explosives, light, and sound.

Whether a machine tool is simple or complex, it performs one or more of these operations. Variations of the six basic operations are employed to meet special situations. For example, there are machine tools that combine two or more operations, as in a boring, drilling, and milling machine; a stamping, punching, and shearing press; or a combination milling and planing machine.

Drilling is a basic machine shop operation dating back to primi-

Fig. 2-2. The automatic tool changer on the vertical spindle machining center. Tools are stored in pockets connected to an endless chain conveyor.

Fig. 2-3. Tool in cut, quill down, stored tools in covered enclosure.

Fig. 2-4. Quill up; fingers grasp the adapter; cover opens, exposing the tool storage conveyor.

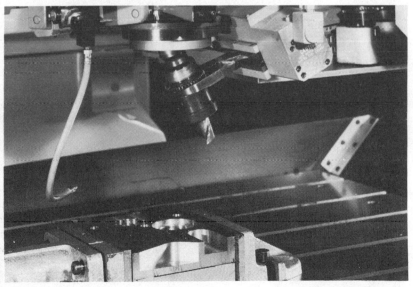

Fig. 2-5. Tool leaves the spindle.

Fig. 2-6. The tool swings in the arc.

Fig. 2-7. The tool is in storage.

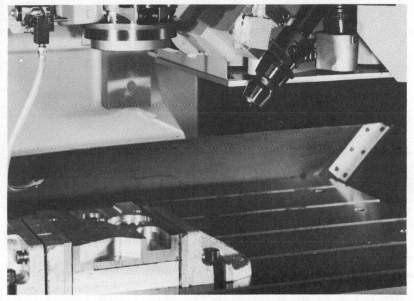

Courtesy Monarch Cortland

Fig. 2-8. The next tool starts to spindle.

tive man. It consists of cutting a round hole by means of a rotating drill. See Fig. 2-9. Boring, on the other hand, involves the finishing of a hole already drilled or cored. This is accomplished by means of a rotating, offset, single-point tool that resembles somewhat the tool used in a lathe or a planer. The tool is stationary and the work revolves on some boring machines. On other types of boring machines, the reverse is true.

Two other types of machine tools are included under the classification of drilling and boring. These machine tools perform reaming and tapping operations. Reaming consists of finishing an already drilled hole. This is done to very close tolerances. Tapping is the process of cutting a thread inside a hole so that a screw may be used in it.

Turning is done on a lathe. The lathe, as the turning machine is commonly called, is the father of all machine tools. The principle of turning has been known since the dawn of civilization, probably originating as the potter's wheel. In the turning operation, the piece of metal to be machined is rotated and the cutting tool is advanced against it. See Fig. 2-10.

41

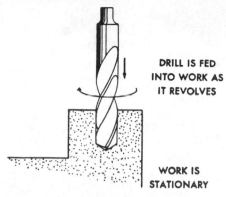

DRILL IS FED
INTO WORK AS
IT REVOLVES

Fig. 2-9. Drilling.

WORK IS
STATIONARY

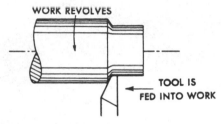

WORK REVOLVES

Fig. 2-10. Turning.

TOOL IS
FED INTO WORK

By contrast, the turret lathe is a lathe equipped with a multisided toolholder called a turret, to which a number of different cutting tools are attached. This device makes it possible to bring several different cutting tools into successive use and to repeat the sequence of machining operations over and over again without the need to reset the tools. The cutting tools themselves are mounted and protrude from the turrets.

Single- and multiple-spindle automatics are used when the number of identical parts to be turned is increased from a few to hundreds or even thousands. These machines perform as many as six or eight different operations at one time on a number of different parts. Single- and multiple-spindle automatics are entirely automatic. Once set up and put into operation, these machines relieve the operator of all but two duties. The duties the operator must do consist of monitoring operations and gaging the accuracy of finished parts.

Milling consists of machining a piece of metal by bringing it into contact with a rotating cutting tool with multiple cutting edges. See Fig. 2-11. A narrow milling cutter resembles a circular saw

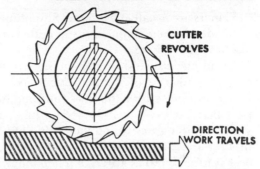

Fig. 2-11. Milling.

CUTTER
REVOLVES

DIRECTION
WORK TRAVELS

blade familiar to most people. Other milling cutters may have spiral edges, which give the cutter the appearance of a huge screw.

There are many other milling machines designed for various kinds of work. The planer type, for example, is built like a planer; but it has multiple-tooth revolving cutters. Machines that use the milling principle but are built especially to make gears are called hobbing machines. Some of the shapes produced on milling machines are extremely simple, like the slots and flat surfaces produced by circular saws. Other shapes are more complex and may consist of a variety of combinations of flat and curved surfaces, depending upon the shape of the cutting edges of the tool and the path of travel of the tool.

Planing or shaping metal with a machine tool is a process somewhat similar to planing wood with a carpenter's hand plane. The essential difference lies in the large size of the machine tool and the fact that it is not portable. The cutting tool remains in a fixed position, while the work is moved back and forth beneath it.

On a shaper the process is reversed. The workpiece is held stationary; the cutting tool travels back and forth. See Fig. 2-12.

A somewhat similar operation is known as slotting. This operation is performed vertically. Slotters, or vertical shapers, are used principally to cut certain types of gears.

Broaches may be classified as planing machines. The broach has a number of cutting teeth; each cutting edge is a little higher than the previous one, and it is graduated to the final size required. The broach is pulled or pushed over the surface to be finished. It may be applied internally, for example, to finish a square hole; or externally to produce a flat surface or a special shape.

43

Planers are usually very large. Sometimes they are large enough to handle the machining of surfaces that are 15 to 20 feet wide and about twice as long.

Grinding consists of shaping a piece of work by bringing it into contact with a revolving abrasive wheel. See Fig. 2-13. The process is often used for the final finishing to close dimensions of a part that has been heat-treated to make it very hard. In recent years, grinding has found increasing applications in heavy-duty metal removal, replacing machines with cutting tools. This process is referred to as abrasive machining.

The grinding machine can correct distortions that have resulted from the heat treatment process. It may be used on external cylindrical surfaces, in holes, for flat surfaces, and for threads.

Under the classification of grinding are included operations known as lapping and honing.

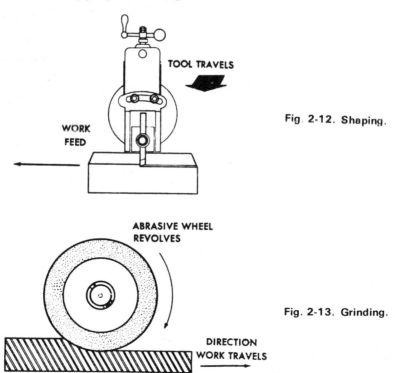

TOOL TRAVELS

WORK
FEED

Fig. 2-12. Shaping.

ABRASIVE WHEEL
REVOLVES

Fig. 2-13. Grinding.

DIRECTION
WORK TRAVELS

Lapping involves the use of abrasive pastes and compounds. It is limited in its use to extremely small amounts of stock removal and to situations where there is a high degree of precision and surface finish needed.

The honing technique, in contrast, is widely accepted as a process separate and apart from that of lapping. For example, there are honing machines with rotating heads that carry abrasive inserts for the extremely accurate finishing of holes.

Metal forming includes shearing, stamping, pressing, and forging metals of many kinds. It requires the use of many kinds of tools. Basically, they are:

1. The shear. This tool is used to cut metal into the required shapes.
2. The punch press. This tool is used to punch holes in metal sheet and plate.
3. The mechanical press. This tool is used to blank out the desired shape from a metal sheet and squeeze it into the final shape in a die under tremendous pressure.
4. The hydraulic press. This tool does the same work as the mechanical press by the application of hydraulic power.
5. The drop hammer. This tool is operated by steam or air. It is used to forge or hammer white-hot metal on an anvil.
6. The forging machine. This tool squeezes a piece of white-hot metal under great pressure in a die. During the process, the metal flows into every part of the die cavity where it assumes the shape of the cavity.

The research and development efforts of the past few years have resulted in a number of new operations for shaping metal into useful parts. Discounting entirely those processes that form metal in its molten state or in a powder state, many new developments have the effect of broadening the capabilities of machine tools.

Some of the better known developments are:

1. Abrasive Machining
2. Capacitor Discharge Machining
3. Cold Extrusion
4. Combustion Machining
5. Electrical Discharge Machining

6. Electromechanical Machining
7. Electrolytic Machining
8. Electrospark Forming
9. Electron Beam Machining
10. Explosive forming
11. Fission Fragment Exposure and Etching
12. Gas Forming
13. Hot Machining
14. Hydroforming
15. Laser (light beam) Cutting
16. Magnetic forming
17. Plasma Machining
18. Spark Forming
19. Ultrasonic Cutting and Forming

A number of the processes were developed to do specific work, such as machining extremely hard materials. Many of the developments came about through aircraft, atomic energy, and rocket research. Some of these processes have been adapted to production-type machines, as in electrochemical milling, electrical discharge machining, magnetic forming, abrasive machining, electron beam machining, electrospark forming, and cold extrusion. A number of processes are still awaiting the basic research necessary to incorporate a process into a piece of production equipment. Although some of these processes are today's curiosities, they will play an important part in manufacturing goods for a rapidly growing economy.

SUMMARY

Machines tools are relatively new. The machine tools in use today are the products of the Industrial Revolution that started a little more than two hundred years ago.

Machine power was substituted for muscle power to cut, shape, and form metal products. Drilling, turning, milling, planing, grinding, and metal forming, the basic machine tool processes in use today, evolved from hand tools.

Drilling is the basic machine shop operation and dates back to primitive man. It consists of cutting a round hole by means of a rotating drill.

Turning is done on a lathe. The lathe, as the turning machine is commonly called, is the father of all machine tools. In the turning operation, the piece of metal to be machined is rotated and the cutting tool is advanced against it.

Milling consists of machining a piece of metal by bringing it into contact with a rotating cutting tool with multiple cutting edges. There are many different types of milling machines.

Planing or shaping metal with a machine tool is a process somewhat similar to planing wood. The cutting tool remains in a fixed position while the work is moved back and forth beneath it.

Grinding consists of shaping a piece of work by bringing it into contact with a revolving abrasive wheel. Grinding can correct distortions that have resulted from the heat treatment process. It may be used on external cylindrical surfaces, in holes, for flat surfaces, and for threads.

Metal forming includes shearing, stamping, pressing, and forging metals of many kinds. It requires the use of many kinds of tools. Basically, they are: the shear, the punch press, the mechanical press, the hydraulic press, the drop hammer, and the forging machine.

The research and development efforts of the past few years have resulted in a number of new operations for shaping metal into useful parts. These new operations will be used more and more in the years ahead.

REVIEW QUESTIONS

1. What is drilling?
2. What is turning?
3. What is milling?
4. What is planing?
5. What is grinding?
6. What is metal forming?
7. What is meant by the machine tool industry?
8. What is a spindle?

9. What are the classifications of machine tools as grouped by the operations they perform?
10. How can turning be done on a lathe?
11. How is a turret lathe different from a machine lathe?
12. How does the planer work?
13. What is a shaper?
14. How are broaches classified?
15. Describe lapping and honing.

CHAPTER 3

Drilling Machines

Round holes are commonly drilled in metal by means of a machine tool called a *drill press*. The term drilling machines is much broader in meaning and includes all types of machines designed for drilling holes into metal.

Many operations other than drilling a round hole can be performed on the drill press. Some of these are sanding, counterboring and countersinking, honing, reaming, lapping, and tapping. Considerable skill is required to drill a hole of proper size in exactly the desired location at a high rate of production. The machine operator must be able to locate the hole properly and accurately, and he must be able to align the drill correctly.

CONSTRUCTION

Successful operation of the drill press requires the operator to be familiar with all parts, and he must have an adequate working knowledge of the machine itself. The operator must also be able to set up the work properly, to select proper speeds and feeds, and to use the correct coolant.

Parts of the Drill Press

The bench-type drill press (Fig. 3-1) and the floor-type drill press (Fig. 3-2) are commonly found in home workshops and in

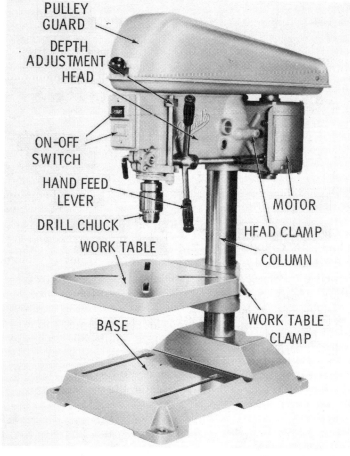

PULLEY
GUARD

DEPTH
ADJUSTMENT
HEAD

ON-OFF
SWITCH

HAND FEED
LEVER

DRILL CHUCK

WORK TABLE

BASE

MOTOR

HFAD CLAMP

COLUMN

WORK TABLE
CLAMP

Fig. 3-1. Bench-type sensitive drill press.

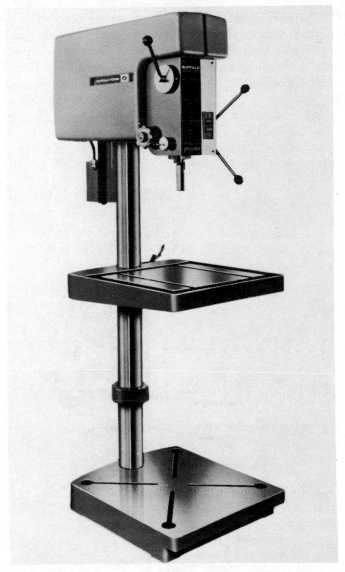

Fig. 3-2. Floor-type sensitive drill press.

industrial machine shops. These machines are designed to rotate a cutting tool (twist drill, countersink, counterbore, etc.) to advance the cutting tool into the metal and to support the workpiece.

Head—The design of the drill press head varies with different machines (Fig. 3-3). In most machines, the electric motor is bolted to the head, and a V-belt drive is used to drive the spindle at from three to five different speeds by shifting the V-belt from step-to-step on the pulleys. For maximum life of the V-belt, the belt tension should be just tight enough that loosening is not necessary to shift it.

Courtesy Buffalo Forge

Fig. 3-3. View of a drill press head with the guard in the open position.

Spindle—The spindle is the rotating part. It is usually splined and made of alloy steel. The spindle rotates and moves up and down in a quill or sleeve, which slides on bearings. A pinion engages a rack fastened to the quill to provide vertical movement of the quill, permitting the twist drill to be either fed into or withdrawn from the workpiece. A typical spindle assembly is shown in Fig. 3-4.

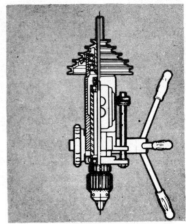

Fig. 3-4. A typical spindle assembly for a drill press.

Courtesy Buffalo Forge

Spindle speed is controlled on smaller machines by changing the V-belt from one step to another on the pulleys. Gear boxes are provided on the larger machines for making changes in spindle speed.

Some upright drill presses have *back gears*, which supply more power to the spindle. Slower spindle speeds are a result of this increase in power. On some machines the back gears can be engaged and disengaged by means of a lever conveniently located on the machine.

Table—The table is supported on the column of the drill press. It can be moved both vertically and horizontally to the desired working position, or it can be swung around so that it will be out of the way. Most tables are slotted so that the work, or a drill press vise for holding the work, can be bolted to them (Fig. 3-5).

Base—the supporting member of the entire drill press structure is the base. It is a heavy casting with holes or slots for bolting it to the bench and for securing the work or workpiece directly to the

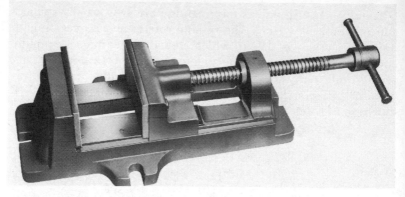

Fig. 3-5. A drill press vise used for holding the work on the drill press table.

base. The base supports the column, which in turn supports the table and head.

Feed—The feed on a drill press can be either manual or automatic. Feed is expressed in thousandths of an inch per revolution of the spindle. The feed of a twist drill is the distance the drill moves into the work per revolution of the spindle. Automatic feed is always referred to as power feed. A hand lever or handwheel is used on manual-type drill presses in which small twist drills are mounted for light work.

Capacity—Several ways are used to express capacity of a drill press. Usually, drill presses are rated by the distance from the center of the twist drill to the column of the machine (Fig. 3-6). Some drill presses are rated by the distance from the top of the table in its lowest position to the tip of the spindle at its highest position. Another rating is given as the distance the spindle can travel from its uppermost position to its lowest position. Still another rating is the largest straight shank twist drill that can be mounted in the drill press spindle or drill chuck.

TYPES OF DRILLING MACHINES

Several types of drilling machines are in existence. Manual-feed drill presses are either light-duty or medium-duty machines. Those with automatic or power feed are heavy-duty machines.

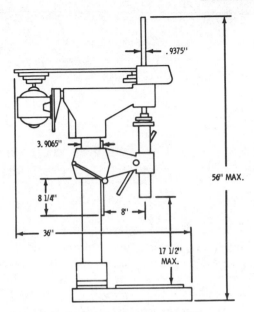

Fig. 3-6. Note the capacity of the bench-type sensitive drill.

Drilling machines are sometimes classified as either vertical-spindle or horizontal-spindle machines and as either single-spindle or multi-spindle machines. The multi-spindle machines are also called gang drilling machines.

Sensitive Drill Press

These machines can be of either the bench type (see Fig. 3-1) or the floor type (see Fig. 3-2). They are belt-driven, hand-fed drill presses. The hand feed of the sensitive drill press permits the operator to "feel" the cutting action at the end of the twist drill. A counterbalanced spindle is moved vertically with a hand lever or handwheel. These machines are designed for relatively light work with small twist drills which are liable to breakage under power feed.

Twist drills up to ½ inch in diameter can be used on the sensitive drill press. The end of the spindle is bored for a standard No. 2 Morse taper, to fit the tapered shank of a drill chuck or a twist drill.

Back-Geared Upright Drill Press

The back-geared drill press (Fig. 3-7) has a greater range of speeds than the floor-type sensitive drill press. It is also larger and equipped with a power feed. The reversing mechanism on the back-geared upright drill press also permits tapping operations.

Radial Drill Press

In a radial drill press (Fig. 3-8), the vertical spindle can be positioned horizontally and locked on an arm that can be swiveled about, and raised and lowered on a vertical column. Thus, the spindle can be placed in any position within its range. The various arm and spindle movements are shown in Fig. 3-9. Some radial drills do not have the last two movements illustrated.

As the drilling head is moveable to any position, it is not necessary to move heavy work for each hole that is to be drilled; therefore, the radial drill is especially adaptable for work of this kind. The radial drill is a heavy-duty drilling machine; it is capable of handling work that, because of its weight or size, cannot be mounted on a drill press table.

A radial drill can be equipped with either a plain table or a universal tilting table (Fig. 3-10). The universal tilting table is designed for angular drilling. The table can be rotated through 360°. When several drills are mounted on the same table, the machine is called a gang drilling machine (Fig. 3-11).

SUMMARY

A drill press is a machine commonly used to drill round holes in metal or wood. Many operations other than drilling round holes can be performed on a drill press. Some of these are sanding, counterboring, countersinking, honing, reaming, and tapping.

Various parts of the drill press are the head, spindle, table, base, and feed. The head generally houses the drill motor and the V-belt drive assembly. The spindle is the rotating part of the drill and slides up and down in a sleeve, which slides on bearing. The spindle also holds the drill chuck, which permits the twist drill to be either fed or withdrawn from the workpiece. The table is

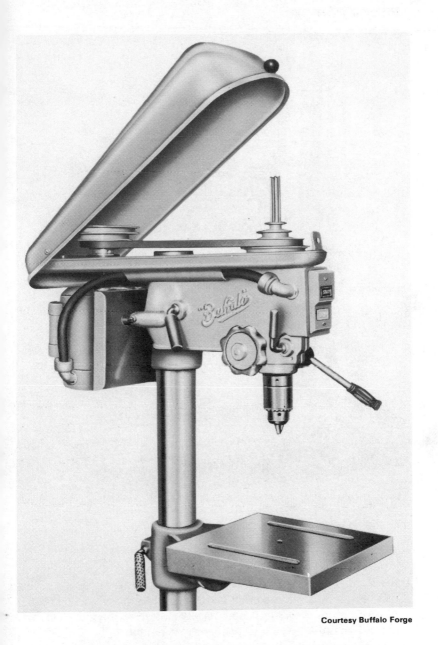

Fig. 3-7. A back-geared motor drive gives a greater range of speeds.

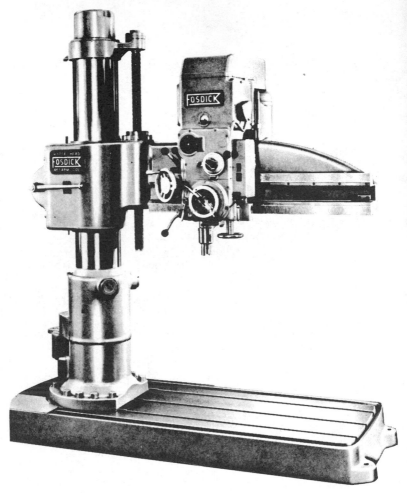

Fig. 3-8. A radial drilling machine.

Courtesy Buffalo Forge

supported on a column of the drill press. The table can be moved both vertically and horizontally to the desired working position, or it can be swung around so that it will be out of the way. The table also supports the work and is generally slotted so the work can be clamped or bolted down to prevent movement.

The feed can be either manual or automatic. Automatic feed is

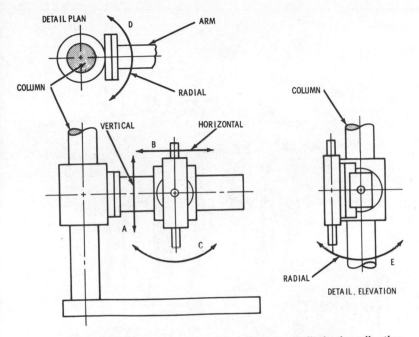

Fig. 3-9. Note the movements of a radial drill having unlimited application or motion.

Courtesy Giddings and Lewis Machine Tool Co.

Fig. 3-10. Radial drill press tables. The plain table is shown (left). The universal or tilting table (right) can be rotated through 360° and tilted through 90°.

Fig. 3-11. A bench-type, six-spindle gang drilling machine.

always referred to as power feed. The feed of a drill is the distance the drill moves into the work.

REVIEW QUESTIONS

1. Explain the purpose of the drill press.
2. Name the basic parts of a drill press.
3. What is meant by the capacity of a drill press?
4. What is a radial drill press?
5. How is spindle speed controlled in a drill press?
6. Why is spindle speed important in a drill press?
7. Why do you need a drill press vise?
8. What is the advantage of a back-geared motor drive on a drill press?
9. What is the advantage of a six-spindle drill press?
10. Where are six-spindle drill presses used?

CHAPTER 4

Drilling Machine Operations

Many operations can be performed on the drilling machine. The setup for any drilling machine operation should be carefully studied and checked before proceeding with the operation.

TYPES OF OPERATIONS

Good sound judgement and trial and error are important in each drill press operation. Hard and fast rules are not practical for operation of the drill press because composition and hardness of material, type of machine, condition of the machine, condition of the cutting tool, lubricant, depth of hole, and many other factors influence the speed and feed at which a meterial can be worked. However, suggestions can be used as a guide, and the operator can make intelligent observations and adjustment of feeds and speeds for a given operator.

Drilling

The chief operation performed on the drill press is drilling. Drilling is the removal of solid metal to form a circular hole. Prior to drilling a hole in metal, the hole is located by drawing two lines at right angles, and a center punch is used to make an indentation for the drill point at the center to aid the drill in getting started. See Fig. 4-1. *Never* attempt to start a twist drill without first using a center punch to make the indentation for starting the drill point.

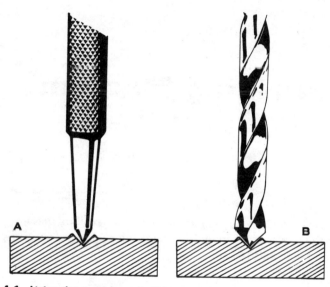

Fig. 4-1. Using the center punch. (A) Center punch used to make an indentation in the workpiece. (B) Indentation should be large enough for the point of the drill.

The Twist Drill

Twist drills are used in drilling machine operations. They are usually made from carbon steel or high-speed steel. Carbon steel twist drills are not suited for high-speed production work. They have to be operated at lower cutting speeds than twist drills made from high-speed steel.

All twist drills have three major parts: the shank, the body, and the point. See Fig. 4-2. They are available in many different sizes;

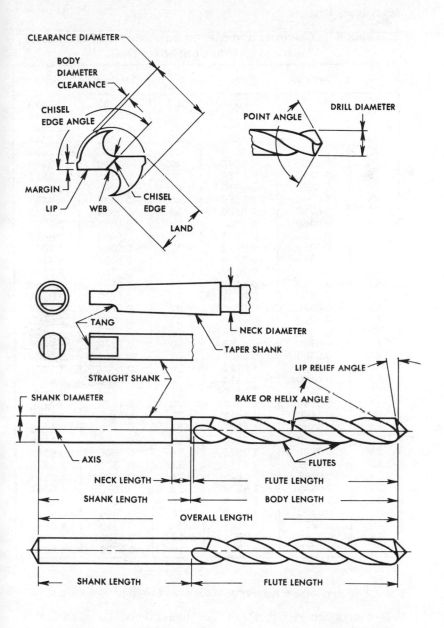

Fig. 4-2. Elements of twist drills.

Courtesy Aluminum Co. of America

Table 4-1. Decimal and Metric Equivalents of Number, Letter, and Fractional Size Drills

Drill Size	Decimal	mm	Drill Size	Decimal	mm
80	0.0135	0.0.3429	47	0.0785	1.9939
79	0.0145	0.3683	46	0.0810	2.0574
1/64	0.0156	0.3969	45	0.0820	2.0828
78	0.0160	0.4064	44	0.0860	2.1844
77	0.0180	0.4572	43	0.0890	2.2606
76	0.0200	0.5040	42	0.0935	2.3749
75	0.0210	0.5334	3/32	0.0938	2.3812
74	0.0224	0.5715	41	0.0960	2.4384
73	0.0240	0.6096	40	0.0980	2.4892
72	0.0250	0.6350	39	0.0995	2.5273
71	0.0260	0.6604	38	0.1015	2.5781
70	0.0280	0.7072	37	0.1040	2.0410
69	0.0292	0.7417	36	0.1065	2.7051
68	0.0310	0.7874	7/64	0.1094	2.7781
1/32	0.0313	0.7937	35	0.1100	2.7940
67	0.0320	0.8128	34	0.1110	2.8194
66	0.0330	0.8382	33	0.1130	2.8702
65	0.0350	0.8890	32	0.1160	2.9464
64	0.0360	0.9144	31	0.1200	3.0480
63	0.0370	0.9398	1/8	0.1250	3.1750
62	0.0380	0.9652	30	0.1285	3.2659
61	0.0390	0.9906	29	0.1360	3.4544
60	0.0400	1.0160	28	0.1405	3.5687
59	0.0410	1.0414	9/64	0.1406	3.5719
58	0.0420	1.0668	27	0.1440	3.6596
57	0.0430	1.0922	26	0.1470	3.7338
56	0.0465	1.1811	25	0.1495	3.7773
3/64	0.0469	1.1906	24	0.1520	3.8608
55	0.0520	1.3208	23	0.1540	3.9116
54	0.0550	1.3970	5/32	0.1562	3.9687
53	0.0595	1.5113	22	0.1570	3.9878
1/16	0.0625	1.5825	21	0.1590	4.0386
52	0.0635	1.6129	20	0.1610	4.0894
51	0.0670	1.7018	19	0.1660	4.2164
50	0.0700	1.7780	18	0.1695	4.3053
49	0.0730	1.8542	11/64	0.1719	4.3656
48	0.0760	1.9304	17	0.1730	4.3942
5/64	0.0781	1.9844	16	0.1770	4.4958

see Table 4-1. These different sizes are classified as follows:

1. Fractional: 1/64″ to 4″ × 1/64″ increments
2. Millimeter
3. Number: 1 to 60, 61 to 80
4. Letter: A to Z

Table 4-1. Decimal and Metric Equivalents of Number, Letter, and Fractional Size Drills (Continued)

Drill Size	Decimal	mm	Drill Size	Decimal	mm
15	0.1800	4.5720	$^{21}/_{64}$	0.3281	8.3344
14	0.1820	4.6228	Q	0.3320	8.4328
13	0.1850	4.6990	R	0.3390	8.6106
$^3/_{16}$	0.1875	4.7625	$^{11}/_{32}$	0.3438	8.7312
12	0.1890	4.8006	S	0.3480	8.8392
11	0.1910	4.8514	T	0.3580	9.0932
10	0.1935	4.8149	$^{23}/_{64}$	0.3594	9.1281
9	0.1960	4.9784	U	0.3680	9.3472
8	0.1990	5.0546	$^3/_8$	0.3750	9.5250
7	0.2010	5.1054	V	0.3770	9.5758
$^{13}/_{64}$	0.2031	5.1594	W	0.3860	9.8044
6	0.2040	5.1816	$^{25}/_{64}$	0.3906	9.9219
5	0.2055	5.2197	X	0.3970	10.0380
4	0.2090	5.3086	Y	0.4040	10.2616
3	0.2130	5.4102	$^{13}/_{32}$	0.4062	10.3187
$^7/_{32}$	0.2188	5.5562	Z	0.4130	10.4902
2	0.2210	5.6134	$^{27}/_{64}$	0.4219	10.7156
1	0.2280	5.8012	$^7/_{16}$	0.4375	11.1125
A	0.2340	5.9436	$^{29}/_{64}$	0.4531	11.5094
$^{15}/_{64}$	0.2344	5.9531	$^{15}/_{32}$	0.4688	11.9062
B	0.2380	6.0452	$^{31}/_{64}$	0.4844	12.3031
C	0.2420	6.1468	$^1/_2$	0.5000	12.7000
D	0.2460	6.2484	$^{33}/_{64}$	0.5156	13.0969
E $^1/_4$	0.2500	6.3500	$^{17}/_{32}$	0.5313	13.4937
F	0.2570	6.5278	$^{35}/_{64}$	0.5489	13.8900
G	0.2610	6.6294	$^9/_{16}$	0.5625	14.2875
$^{17}/_{64}$	0.2656	6.7469	$^{37}/_{64}$	0.5781	14.6844
H	0.2660	6.7564	$^{19}/_{32}$	0.5938	15.0812
I	0.2720	6.9088	$^{39}/_{64}$	0.6094	15.4781
J	0.2770	7.0358	$^5/_8$	0.6250	15.8750
K	0.2810	7.1374	$^{41}/_{64}$	0.6406	16.2719
$^9/_{32}$	0.2812	7.1437	$^{21}/_{32}$	0.6562	16.6687
L	0.2900	7.3660	$^{43}/_{64}$	0.6719	17.0656
M	0.2950	7.4930	$^{11}/_{16}$	0.6875	17.4625
$^{19}/_{64}$	0.2969	7.5406	$^{45}/_{64}$	0.7031	17.8594
N	0.3020	7.6508	$^{23}/_{32}$	0.7188	18.2562
$^5/_{16}$	0.3125	7.9375	$^{47}/_{64}$	0.7344	18.6531
O	0.3160	8.0264	$^3/_4$	0.7500	19.0500
P	0.3230	8.2042	$^{49}/_{64}$	0.7656	19.4469

There are many different kinds of twist drills. Some twist drills are designed for specific types of work. A number of different twist drills are illustrated, with their specific applications, in Figs. 4-3 through 4-13.

Before any twist drill is used, care must be taken to see that the

Table 4-1. Decimal and Metric Equivalents of Number, Letter, and Fractional Size Drills (Continued)

Drill Size	Decimal	mm	Drill Size	Decimal	mm
$25/32$	0.7812	19.8437	$29/32$	0.9062	23.0187
$51/64$	0.7969	20.2406	$59/64$	0.9219	23.4156
$13/16$	0.8125	20.6375	$15/16$	0.9375	23.8125
$53/64$	0.8281	21.0344	$61/64$	0.9531	24.2094
$27/32$	0.8438	21.4312	$32/32$	0.9688	24.6062
$55/64$	0.8594	21.8281	$63/64$	0.9844	25.0031
$7/8$	0.8750	22.2250	1	1.0000	25.4001
$57/64$	0.8906	22.6219			

twist drill is properly ground. Figs. 4-14 through 4-19 illustrate the correct and incorrect procedures.

Drilling Suggestions—The operator should check to make certain that the drill point has started properly before the drill has cut too deeply. Frequently, the drill point fails to seat properly in the punch mark, and a small hole is made off center.

The operator should use layout lines and layout circles so that he can determine whether the twist drill is making a hole concentric with the layout circle at the start and is going properly. If the small hole made by the drill is not concentric with the layout circle, the drill should be withdrawn. A cape chisel can be used to cut a groove, or several grooves, on the side toward which the drill should be moved so that the drill can be "drawn" back to the center of the layout circle. It is too late to shift the twist drill point without marring the workpiece, after the drill has begun to cut its full diameter.

After the mounted work is placed on the worktable of the drill press, the worktable should be adjusted to a convenient height for drilling, considering both the length of the twist drill and drill-holder. The height of the spindle should be adjusted to provide the shortest distance for feed of the twist drill to the work.

After the proper speed has been determined, the machine can be started, and the twist drill lowered to the work. The drill should be fed to the work slowly until the metal has been penetrated by the point of the drill. Then a check should be made to see that the drill has started properly.

Feed and speed—In drilling holes with a small diameter, the danger of twist drill breakage is very great unless the feed and

Fig. 4-3. Taper-length, tanged, automotive series, straight-shank twist drill regularly regularly furnished with tangs for use with split-sleeve drill drivers.

Fig. 4-4. Straight-shank, three-flute core drill.

Fig. 4-5. Straight-shank, four-flute core drill.

Fig. 4-6. Straight-flute drill for free machining brass, bronze, or other soft materials, particularly on screw machines. Also suitable for drilling thin sheet material, because of lack of tendancy to 'hog."

Fig. 4-7. Drill used principally for drilling molded plastics. Also used for drilling hard rubber, wood, and aluminum and magnesium alloys.

Fig. 4-8. Low-helix drill used extensively in screw machines making parts from screw stock and brass.

Fig. 4-9. High-helix drill designed with higher cutting rake and improved chip conveying properties for use on materials such as aluminum, die-casting alloys, and some plastics.

Fig. 4-10. Screw-machine length straight-shank twist drill.

Fig. 4-11. Center drill.

Fig. 4-12. Starting drill.

Fig. 4-13. Oilhole twist drills for production work in all types of materials on screw machine or turret lathes.

Fig. 4-14. The two cutting edges (lips) should be equal in length, and should form equal angles with the axis of the drill. Angle C should be 135° for drill hard or alloy steels. For drilling soft materials and for general purposes, angle C should be 118°.

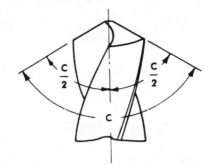

Courtesy National Drill and Tool Co.

Fig. 4-15. A twist drill with the lips ground at unequal angles with the axis of the drill can be the cause of an oversized hole. Unequal angles also result in unnecessary breakage and cause the drill to dull quickly.

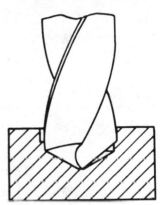

Courtesy National Drill and Tool Co.

Fig. 4-16. The result of grinding the drill with equal angles, but having the lips unequal in length.

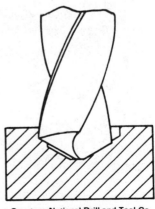

Courtesy National Drill and Tool Co.

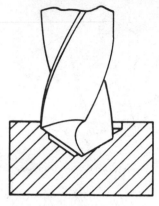

Fig. 4-17. The result of grinding the drill with lips of unequal angles, and with lips of unequal lengths.

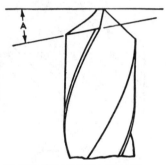

Fig. 4-18. The lip relief angle A should vary according to the material to be drilled, and according to the diameter of the drill. Lesser relief angles are required for hard and tough materials than for soft, free-machining materials.

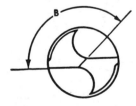

Fig. 4-19. The chisel-point angle B increases or decreases with the relief angle and should range from 115° to 135°.

Courtesy Giddings & Lewis Machine Tool Co.

Fig. 4-20. Performing small-hole drilling on the Bickford radial drilling machine. Drills and taps as small as $3/16$ inch (0.1875) and $1/4$ (0.250) can be handled at high speeds.

speed are given careful attention, especially at the moment the point of the drill breaks through the other side of the work (Fig. 4-20). In general, high speed and light feed are recommended. It is better to err on the side of too much speed than to err on the side of too much feed (except for cast iron, which permits an unusually heavy feed). Speed can be increased to the point where the outside corners of the twist drill show signs of wearing away. Speed can then be reduced and maintained at the reduced speed. Recommended drilling speeds can be obtained from Table 4-2.

Too much speed is indicated if the twist drill chips out at the cutting edge (Table 4-3). This is a certain indication of either too heavy feed or too much lip clearance (Fig. 4-21). If the twist drill splits up the web, too much feed for the amount of lip clearance is indicated. Either the feed should be decreased or the lip clearance should be increased—or both actions might be used to remedy the difficulty. Insufficient lip clearance can also cause a twist drill to split up the web. Also, too much lip clearance at the center (or at

Table 4-2. Recommended Drilling Speeds for Materials
(High-Speed Drills)

Material	Recommended Speed in Surface Feet per Minute (sfm)
Aluminum and alloys, Brass and bronze, soft ...	200-300
Bakelite	100-150
Plastics.......................................	100-150
Bronze, high tensile	70-100
Cast iron, chilled, Steel, stainless (hard)	30-40
Cast iron, hard	70-100
Cast iron, soft	100-150
Magnesium and alloys	250-400
Malleable iron	80-90
Monel, metal	40-50
Nickel	40-60
Steel, annealed (.4 to .5 percent C)	60-70
Steel, forgings, Wrought iron, Steel, tool	50-60
Steel, machine (.2 to .3 percent C)	80-110
Steel, manganese (15 percent Mn), Slate, marble	15-25
Steel, soft	80-100
Steel, stainless (free machining)	60-70
Wood	300-400

any other point on the lip) can cause the cutting lips to chip. Therefore, if the twist drill is properly ground, decrease the feed to eliminate these conditions.

Precision drilling is required when two or more holes have been laid out in specific relationship to each other and the holes must be drilled accurately to layout marks. Templates, jigs, and fixtures are all used in precision drilling. This also includes drilling holes to a specified depth.

Drilling on the Automatic Machines

High speeds and light feeds are especially recommended for automatic drilling machines where the holes do not exceed four drill diameters in depth. A small compactly rolled chip is desirable. If possible, the chip should be kept unbroken for the entire depth of the hole, as such a chip feeds out through the flutes more easily. If the drill appears to be functioning well, but the surface of the hole is rough, a dull twist drill is indicated, and it should be resharpened.

Courtesy Giddings & Lewis Machine Tool Co.

Fig. 4-21. Using a Bickford radial drilling machine to drill a 1 $\frac{33}{64}$-inch (1.5156) hole in a welded all-steel frame for a three-ton electric hoist. Feed is 0.025 inch per revolution at 175 r/min.

Controls—Most automatic drilling machines are controlled from a handy push button control station. Coolant, spindle jog, spindle drill, and drill-tap operations are controlled by selector switches. Push buttons are used to control cycle start, spindle start, emergency spindle returns, and motor stop functions (Fig. 4-22). The "jog" switch is provided for setups. The "on" position on the "jog" selector switch permits jogging the spindle downward—using the "cycle start" push button, or positioning the spindle with the handwheel on the lower part of the head. The "dwell" switch causes the spindle to pause at the bottom of the feed stroke for a period of time preset by the electric timing switch on the control panel. The "dwell" switch is noneffective in the tapping cycle. The "drill" position on the "drill-tap" selector switch causes the spindle to turn in the forward direction at all times. Set in the "tap" position, the spindle runs in the forward direction at the start of the

75

Fig. 4-22. Control station on Bickford upright drilling ma-
chine, used to control
all machine functions.

Courtesy Giddings & Lewis Machine Tool Co.

cycle, reverses automatically when the bottom limit is reached,
then stops at the top of the travel (see Fig. 4-22).

Depth of feed, upper limit of spindle travel, and start of power
feed may be controlled by limit settings. Dial-type limit settings
are used on some machines (Fig. 4-23). The limit settings are
quickly set on large concentric selector dials. The inner dial limits
the return spindle stroke; the intermediate dial determines the
disengagement of the rapid traverse and instantaneous engagement of the power feed; and the large outer dial determines the
final depth of the tool (see Fig. 4-23).

Automatic cycling—For automatic cycle drilling, the spindle is
advanced to the point set on the intermediate dial by depressing
the "cycle start" button. The spindle advances at power feed to the
depth setting determined by the outer dial. Then, rapid return is
effected to the upper spindle position, determined by the inner
dial, to complete the cycle (see Fig. 4-23).

For automatic cycle tapping, the spindle is rapidly advanced to
the point set on the intermediate dial, and continues at the tap lead
rate to the depth determined by the outer dial. At the bottom limit,

Courtesy Giddings & Lewis Machine Tool Co.

Fig. 4-23. Dial-type limit-setting controls used on the Bickford upright drilling machine.

the spindle reverses, and the tap is retracted at the tap lead rate, until the tap is free of the work and returns to the upper limit at the rapid travel rate; spindle rotation stops, completing the cycle (see Fig. 4-23).

Tapping

The drill press can be used as a means of tapping or cutting threads in a drilled hole. Precision hand tapping can be performed on the drill press with the work mounted on the drill press table in the position as for drilling the hole. A tap wrench is mounted on the square shank of the hand tap, and a lathe center is mounted in the hollow spindle of the drill press. Then the tap is placed in the hole

Table 4-3. Cutting Speeds (Fractional Drills)

Feet per Min.	30′	40′	50′	60′	70′	80′	90′	100′	110′	120′	130′	140′	150′
Diameter, Inches					Revolutions per Minute (r/min)								
1/16	1833	2445	3056	3667	4278	4889	5500	6111	6722	73334	7945	8556	9167
1/8	917	1222	1528	1833	2139	2445	2750	3056	3361	3667	3973	4278	4584
3/16	611	815	1019	1222	1426	1630	1833	2037	2241	2445	2648	2852	3056
1/4	458	611	764	917	1070	1222	1375	1528	1681	1834	1986	2139	2292
5/16	367	489	611	733	856	978	1100	1222	1345	1467	1589	1711	1833
3/8	306	408	509	611	713	815	917	1019	1120	1222	1324	1425	1528
7/16	262	349	437	524	611	698	786	873	960	1048	1135	1224	1310
1/2	229	306	382	459	535	611	688	764	840	917	993	1070	1146
5/8	183	245	306	367	428	489	550	611	672	733	794	857	917
3/4	153	203	254	306	357	407	458	509	560	611	662	713	764
7/8	131	175	219	262	306	349	393	436	480	524	568	613	655
1″	115	153	191	229	267	306	344	382	420	458	497	535	573
1⅛	102	136	170	204	238	272	306	340	373	407	441	476	509
1¼	92	122	153	183	214	244	275	306	336	367	397	428	458
1⅜	83	111	139	167	194	222	250	278	306	333	361	389	417
1½	76	102	127	153	178	204	229	255	280	306	331	357	382
1⅝	70	94	117	141	165	188	212	235	259	282	306	329	353
1¾	65	87	109	131	153	175	196	218	240	262	284	306	327
1⅞	61	81	102	122	143	163	183	204	224	244	265	285	306
2″	57	76	95	115	134	153	172	191	210	229	248	267	287

in the workpiece, and the point of the lathe center is placed in the center hole in the shank end of the tap. The hand-feed lever of the drill press is used to maintain steady pressure without forcing the hand tap, as the tap wrench is used to turn the tap into the work. Actually, the drill press is used, in this instance, chiefly as a guide for precision hand tapping.

A tapping attachment can be used on the drill press. A friction clutch is built into the device. If a tap sticks, jams, or reaches the bottom of a blind hole, the clutch will slip before the tap will break. Raising the spindle of the machine permits a reversing mechanism in the tapping attachment to back out the tap without breaking it.

Automatic drilling machines can be equipped with one of three types of tapping devices:

1. Friction-type tapping attachment.
2. Motor-reverse control operated by a small lever control.
3. Automatic reverse feed handle control.

Fig. 4-24. Motor reversing mechanism used for taps ½ inch to 1 inch in diameter. A magnetic reversing control, actuated by a lever station mounted on the spindle arm, is used to reverse the rotation of the motor.

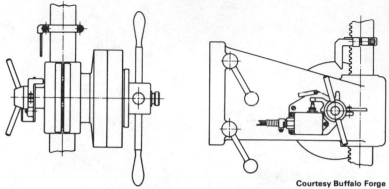

Fig. 4-25. Automatic feed handle reverse tapping control (left) and power feed (right).

Fig. 4-26. Using a Bickford radial drilling machine to tap a 5-inch bore in a motor housing.

Fig. 4-27. Machine countersink. The included angle is 82°.

Fig. 4-28. Straight-shanked machine-screw counterbore.

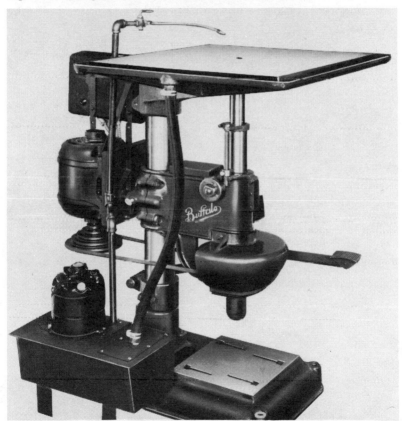

Fig. 4-29. A spot-facing machine built to handle large forgings and castings. This machine is equipped with a coolant system and a foot feed.

The friction-type tapping attachment is usually preferred to motor-reverse tapping for high-production work (Fig. 4-24). Continuous motor reversals are not advisable for continuous operation on machines with back gears and power feed (Fig. 4-25). Also, small taps, up to ⅜ inch, used for high-production work, perform better on the friction-type tapping attachment (see Fig. 4-26).

Countersinking

The operation in which a cone-shaped enlargement is formed at the end of a hole is called countersinking. A conical cutting or reaming tool is used to taper or bevel the end of the hole (Fig. 4-27).

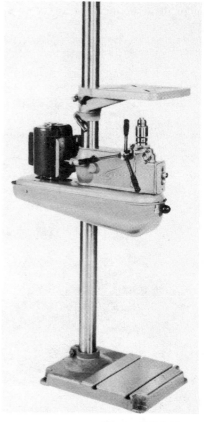

Fig. 4-30. Inverted drill press head for routing or back spot-facing operations on the drill press.

Fig. 4-31. Mortising attachment
for a drill press.

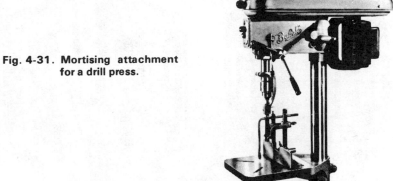

Courtesy Buffalo Forge

Courtesy Giddings and Lewis Machine Tool Co.

Fig. 4-32. Using Bickford radial drilling machine for trepanning, boring, and
tapping operations. Feed of 0.050 inch per revolution is used for
boring and trepanning. Pipe tapping can be performed at 22
r/min.

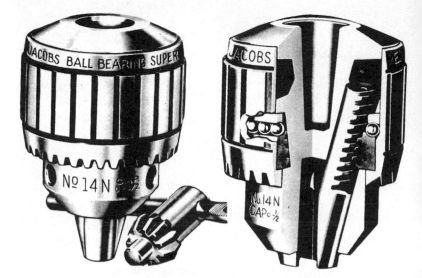

Fig. 4-33. Ball-bearing chuck (left) and cutaway (right), mounted on either straight- or taper-shank arbors for use on drilling machines and other industrial equipment.

The countersinking operation involves fastening the workpiece properly, mounting the countersink in the drill chuck, aligning the countersink with the work, selecting the correct spindle speed, and using care in feeding the countersink to the work. The countersink should be rotated at a relatively slow spindle speed, using cutting oil for a smooth job in steel. Failure to clamp the work properly, a dull tool, or excessive spindle speed can be the cause of a rough countersinking job.

Counterboring

The operation in which a hole is enlarged cylindrically part way along its length is called counterboring. A counterboring tool is used to enlarge the diameter of the hole to accommodate a stud, bolt, or pin having two or more diameters—for example, a filister-head screw.

The counterboring tool (Fig. 4-28) must be properly aligned with the original hole so that the pilot on the end of the counter-

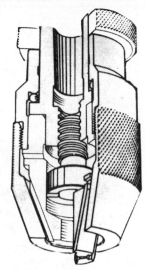

Fig. 4-34. Albrecht keyless chuck (left) and cutaway view (right), used on high-speed sensitive drilling machines and other industrial equipment.

bore will fit into and follow the original hole properly. The size of the enlarged hole is produced by the cutting edges on the counterbore.

Spot-Facing Operation

Another operation similar to counterboring is the spot-facing operation (Fig. 4-29). Just enough metal is removed to provide a bearing surface for a washer, nut, or the head of a cap screw. The spot-facing tool should be mounted in the drill press spindle, and the pilot of the spot-facing tool aligned with the original drilled hole (Fig. 4-30).

Reaming

The reaming operation is used to accurately size and finish the inside of a hole. Reaming is used to make a drilled hole more accurate and more smooth. A hand reamer can be used to finish a hole to the exact dimension. A hole should be drilled 1/64 inch undersize and then machine reamed to obtain an accurate smooth hole of standard size. If the blueprint should call for a ½-inch ream, a twist drill $\frac{1}{64}$ inch under ½ inch should be used to drill the original hole.

Machine reaming can be performed in the drill press by substituting the correct size of reamer for the twist drill in the drill press spindle without changing the mounting position of the work or removing it. The correct spindle speed for machine reaming is

Courtesy Morse Twist Drill and Machine Co.

Fig. 4-35. Morse tapered nylon sleeve. The nylon sleeve is designed to give way under heavy loads before stress can ruin the drill. High resiliency helps prolong life of tools and equipment by absorbing deflections caused by loose spindles, excessive overhand, and dull tools.

Courtesy National Twist Drill and Tool Co.

Fig. 4-36. Steel socket used to hold twist drills with shanks too large to fit into either the drill press spindle or a sleeve.

approximately one-half the drilling speed. The automatic feed can be used for reaming.

The slight taper at the end of a reamer is provided to facilitate entry of the reamer into the hole, and the reamer should be extended through the hole a distance of at least 1½ inches to produce the accurate size. The reamer should never be rotated in the reverse direction, as the edges may be broken or dulled.

Grinding and Buffing

The drill press can be used for simple grinding and buffing operations. The work can be held in a drill press vise with a cup grinding wheel mounted in the chuck.

A burring or polishing operation can be performed by mounting a buffing wheel in the chuck. The work should be held securely.

Fig. 4-37. Drill drift used to remove twist drills, sleeves, and sockets from the drill press spindle.

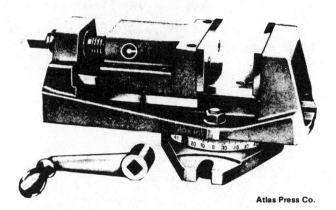

Fig. 4-38. A Clausing machine vise with graduated swivel base.

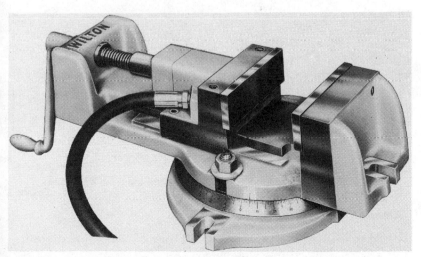

Fig. 4-39. An air-hydraulic vise.

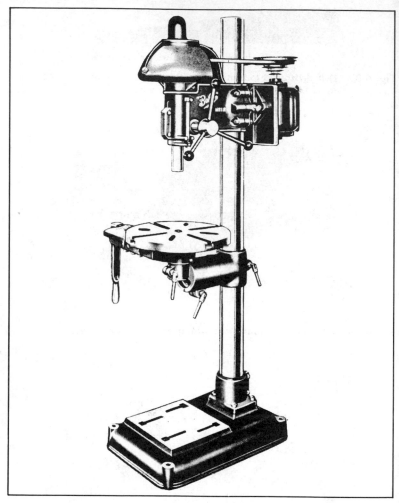

Fig. 4-40. Combination table and vise for drilling machine operations.

Other Operations

Many drill presses can be adapted for *mortising* operations by means of a mortising attachment (Fig. 4-31). The attachment usually fits over the chuck, and the fence and hold down are mounted on the table. A ½-inch-square hole is usually the maxi-

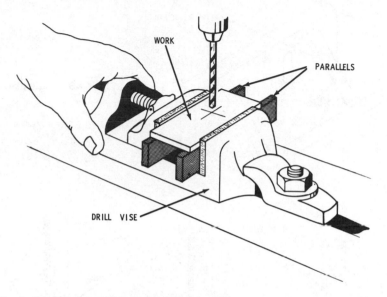

Fig. 4-41. Using parallels in the drill press vise.

Courtesy Lufkin Rule Co.

Fig. 4-42. V-blocks and clamps made in pairs for use where extremely accurate settings are desired.

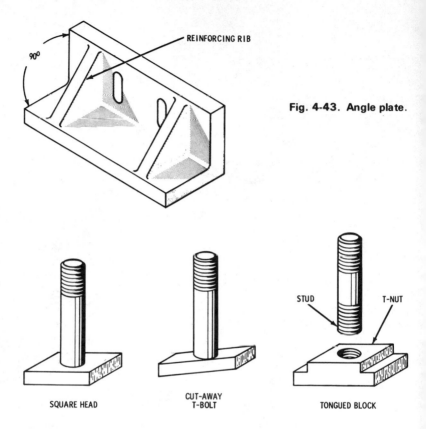

REINFORCING RIB

90°

Fig. 4-43. Angle plate.

STUD

T-NUT

SQUARE HEAD

CUT-AWAY
T-BOLT

TONGUED BLOCK

Fig. 4-44. Common types of T-bolts: (A) square head; (B) cutaway T-bolt; (C) tongued block.

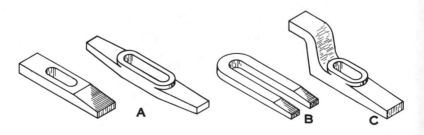

A

B

C

Fig. 4-45. Common types of straps or clamps: (A) flat strap; (B) U-strap; (C) goose-neck strap.

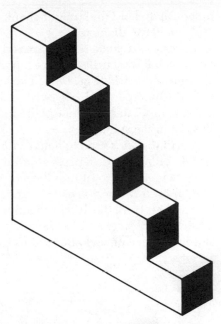

Fig. 4-46. Step block.

mum capacity for a mortising attachment on the drill press. *Trepanning* and *boring operations* can also be performed on the drilling machine (Fig. 4-32).

HOLDING DEVICES

Drill presses and drilling machines require two types of holding devices: (1) devices for holding the tool or cutter; and (2) devices that secure the workpiece in a properly mounted position for drilling, tapping, and so forth.

Tool or Cutter Holders

Most cutting tools have either a straight shank or a taper shank. Most drill presses are equipped with a Morse taper spindle. Straight-shank drills are held in a chuck.

Chucks—Many styles of chucks for a variety of purposes are manufactured (Fig. 4-33 and 4-34). The drill press chuck has a Morse taper on the chuck shank. The tapered shank fits into the drill press spindle. Three small jaws in the chuck tighten simultaneously to hold the straight-shank twist drills, which are ½ inch and smaller in diameter.

Sleeves and sockets—A tapered drill sleeve is used to hold taper-shank twist drills that are too small for the tapered hole in the spindle of the drill (Fig. 4-35). The hole inside the sleeve is tapered to fit the tapered shank of a twist drill. The outside of the sleeve has a Morse taper to fit the hole in the spindle. The tanged end fits into the slot in the spindle.

A drill socket is used to hold twist drills with shanks too large to fit into either the drill press spindle or a sleeve (Fig. 4-36). It also has a tanged end to fit into the slot in the drill press spindle.

A tapered key or *drill drift* is used to remove twist drills, sleeves, and sockets from the drill press spindle (Fig. 4-37). These drifts are made of tool steel and hardened. The drift is driven through the slot in a sleeve or socket to remove a taper-shank twist drill.

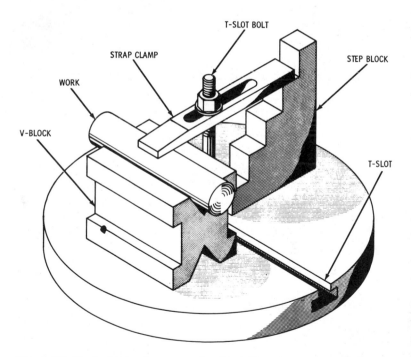

Fig. 4-47. Typical setup in preparation for drilling in round stock.

Work Holders and Setup Devices

The workpiece must be securely fastened to give satisfactory and safe results. Mounting and supporting the workpiece properly are extremely important. Of course, the layout work must be done properly, but the correct procedure for making setups and for operating the drilling machine must be followed carefully for satisfactory results.

Vise—A drill press vise is used to hold and support the work on the drill press table. A vise with a graduated swivel base is shown in Fig. 4-38. An air-hydraulic vise is used on high-production machines (Fig. 4-39). A combination table and vise for drilling machine operations is shown in Fig. 4-40.

Parallels—These accurately machined bars are made in pairs. Parallels are used to raise the work above the drill press table so that the part can be drilled completely through without damage to either the vise or the table. Parallels should be placed carefully so that the twist drill will not damage them after completing its passage through the work (Fig. 4-41).

V-blocks—Round stock can be securely clamped with V-blocks for drilling (Fig. 4-42). Care must be taken and the work clamped securely; otherwise, the operator can be injured when the twist drill takes hold in the workpiece, causing the work to swing around.

Angle plate—The angle plate is useful when it is desirable to drill a hole parallel to another surface. The angle plate is usually made of cast iron, and holes and slots are provided for clamping it to the machine table to secure the work. It is planed on two sides to an angle of exactly 90° (Fig. 4-43).

T-bolts—These bolts are placed in the T-slots provided in the table. Either the workpiece or the vise can be securely fastened to the table (Fig. 4-44).

Straps or clamps—An assortment of straps or clamps can be used to clamp workpieces to the table. The clamps should be made of a good grade of steel to prevent bending under pressure (Fig. 4-45).

Step blocks—A step block is used to support the end of a clamp or strap opposite the work. These blocks are usually made of a good grade of steel and are usually made in pairs (Fig. 4-46). A step block useful as an aid to keeping the strap level when fastened to the work.

Safety

Any movement of either the workpiece or the table can cause the twist drill to break and result in injury to the operator. The workpiece must be clamped securely and rigidly (Fig. 4-47). Bolts should be just long enough and should be placed as near the work as possible. Clamps and blocks should be of correct height and size, and washers should be used on all nuts. Many injuries have been caused by attempting to hold the work with the hands. The operator should always make certain that the work is clamped securely.

SUMMARY

The chief operation performed on the drill press is drilling. Drilling is the removal of solid metal to form a circular hole. Prior to drilling a hole in metal, the hole is located by drawing two lines at right angles, after which a center punch is used to make an indentation for the drill point at the intersection of the two lines to aid in starting the drill.

The operator should check to make certain that the drill point has started properly before the drill has cut too deeply. It is too late to shift the twist drill point (without marring the workpiece) after the drill has begun to cut its full diameter. After the work is placed on the worktable of the drill press, the table should be adjusted to a convenient height for drilling.

Many drill presses can be adapted for mortising operations by means of a mortising attachment. The attachment usually fits over the chuck, and the fence and hold-down are mounted on the table. Trepanning and boring operations can also be performed on the drilling machine.

REVIEW QUESTIONS

1. What operations can be performed on a drill press?
2. What are the three major parts of a twist drill?
3. What precautions should be taken before drilling a hole?
4. What tapping devices are on automatic drilling machines?

CHAPTER 5

The Vertical Boring Mill

The vertical boring mill is virtually equivalent to a lathe turned on end so that the faceplate is horizontal, omitting the tailstock. The table of the boring mill corresponds to the faceplate of the lathe; it is more convenient to clamp heavy work to the horizontal table of the mill than to the vertical faceplate of the lathe.

The machine is adapted to boring and turning heavy work that has a larger diameter than its width or height. As on the lathe, the works turns—the only movement of the tool being due to its feed. More than one tool can be used at a time. The feed is such that the cutting tool can be given horizontal, vertical, or angular movement. Thus the work can be faced parallel with the table, bored, or turned cylindrically or conically.

The work can be either held in a chuck or clamped directly to the table. If the work has an irregular shape or size, clamping it to the table is more desirable; or if a piece is to be machined on both sides, the previously machined surface can be placed on the table for more accuracy. On production work, special fixtures can be used to set up the work more quickly.

CLASSIFICATION

Boring mills are used for various purposes—drilling, reaming, chamfering, counterboring, and so on. They can be classified by the number of rams as either single ram or double ram, and they are available as either single- or double-station units (Fig. 5-1).

Vertical boring mills are easy to load, especially for handling awkward parts. All stations are readily accessible front stations, making it easy for one operator to handle several machines. The vertical mills also save space because their bulk extends upward into unused space, and they can be lined up close to each other.

The size of a boring mill is designated by the size of the table; for example, a 72-inch mill has a table that is 72 inches in diameter. As for the engine lathe, the size designation is called the swing.

Fig. 5-1. A vertical boring machine.

PRINCIPAL PARTS

The basic units of the boring mill are the slide and the boring head. These components can be arranged in almost limitless combinations to provide efficient production line operation.

Slide units in combination with standard wing bases are designed to offer wide adaptability; and they are rigid and stable, regardless of slide position (Fig. 5-2). Slides can be mounted on either the right- or left-hand slide of a transfer line or as multiple operation units.

Courtesy Heald Machine Co.

Fig. 5-2. Slide unit for a vertical boring mill.

The boring head (Fig. 5-3) is designed for rotating and feeding. It can handle almost any drilling, reaming, chamfering, and counterboring operation either as a single unit or in multiple spindle units. A number of the units can be grouped on a single base or on separate bases in an automated line. They can be mounted horizontally, vertically, or at any angle in any plane.

The table of the boring mill (Fig. 5-4) travels on box-type ways. Cross slides, steady rests, or workholding fixtures can be positioned on the table. The table shown in Fig. 5-4 has a constant feed hydraulic system of the locked feed type to maintain constant feed rates.

BORING OPERATIONS

The vertical boring mill operates on the principle that "it is easier to lay down a piece of work than it is to hold it upright," which is probably the chief reason for its development. Three general methods of fastening the work to the table are:

Courtesy Heald Machine Co.

Fig. 5-3. Boring head unit for a boring mill can be used in either single or multiple units.

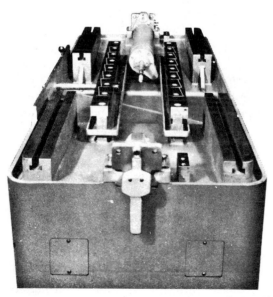

Courtesy Heald Machine Co.

Fig. 5-4. The table of the boring mill is mounted on hardened and ground box-type ways with hold downs. Table travel is accurate even under heavy cuts.

1. Bolts and clamps.
2. Chucks.
3. Fixtures.

Boring

Holes in castings are usually cored, so it is necessary only to finish the rough hole to the required diameter. Usually, a boring tool is used for small holes. Its shank is held in either a ram or a turret. Sometimes a shell drill with four flutes can be used. A very light finishing cut can be taken with a reamer for precision work (Fig. 5-5.)

For large holes, the tool can be held in a tool head or holder on the end of a ram (Fig. 5-6). Several types of boring tools are available. The boring tool can be of a horizontal type or a bent type (see Fig. 5-6). Cast iron is usually finished with a broad, flat tool.

Holding the Work

Properly setting and fastening the work on the table is important in any boring mill operation. If the work to be bored is an irregular piece, it can be fastened to the table by the bolt and clamp method (Figs. 5-7 and 5-8).

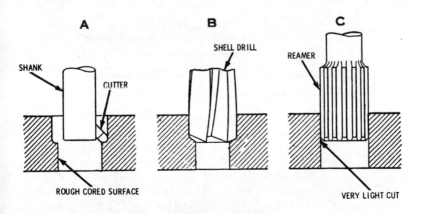

Fig. 5-5. Boring small-cored holes with a shank cutter (A), with a shell drill (B), and finishing with a reamer (C).

101

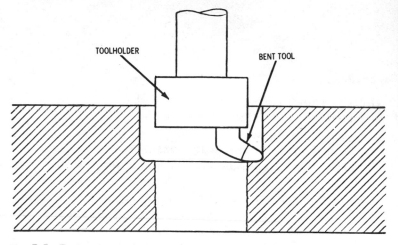

Fig. 5-6. Boring a large hole with the bent tool held in a toolholder.

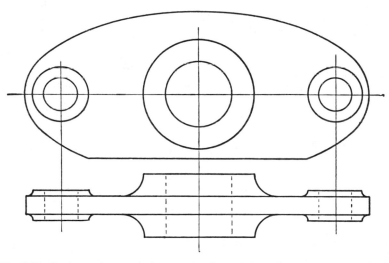

Fig. 5-7. An irregular workpiece to be fastened to the worktable of the boring mill.

The large hole in the work must be centered accurately with the center hole in the table. As the hole to be bored is larger than the hole in the table, the work must be elevated above the table surface to provide clearance for the cutting tool at the lower end of

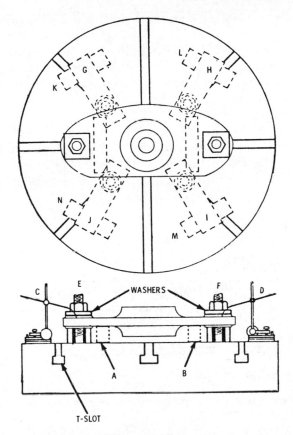

Fig. 5-8. Using bolts and clamps to clamp an irregular workpiece to the worktable.

the bore. This can be accomplished by means of parallel blocks (see A and B in Fig. 5-8), but distance sleeves threaded on the bolts (see E and F in Fig. 5-8) are preferable.

The upper surface of the work should be parallel with the surface of the table. This can be adjusted in several positions by making adjustments with the parallel blocks (see C and D in Fig. 5-8). As the workpiece has a hole at each end, it can easily be fastened to the table by the bolts E and F. The bolts are placed in the T-slots of the table, passed through the end holes in the work, and the nuts turned down firmly onto washers resting on the work.

103

The concentric distance sleeves are preferable to the offset parallel blocks.

To determine whether the hole has been centered properly, fasten a scriber to the toolholder with its point near the circumference of the hole to be bored. Rotate the table slowly, and the scriber will indicate whether the hole has been centered properly for the operation.

If there are no holes at the end of the work, it can be fastened to the table with bolts and clamps (see *G, H, I, J* in Fig. 5-8). The outer end of the flat plate clamps rests on the blocks (see *K, L, M, N* in Fig. 5-8) at the proper height, as shown in detail in Fig. 5-9.

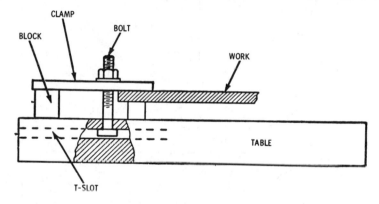

Fig. 5-9. Detail drawing of a workpiece fastened to the worktable by means of bolts and clamps.

Turning

Although the machine is called a vertical boring mill, it can be used more for turning than for boring operations, especially for workpieces that require both turning and boring. The kinds of turning operations that can be performed on the boring mill are: (1) flat (facing); (2) cylindrical; and (3) conical.

Flat and cylindrical turning—A flywheel can be machined on the boring mill to illustrate both flat and cylindrical turning. Fastening the work to the table by chucking can also be illustrated.

Chucks that are built into the table are usually of the combination type, that is, the jaws have both universal and independent adjustment. The chuck jaws should be set against the interior surface of the rim, as the rim can be faced and turned at one setting.

To avoid the risk of the flywheel slipping in the chuck because of the tangential thrust of the cutting tool, place the wheel in the chuck, if possible, so that one spoke will bear against one of the chuck jaws. If this is not feasible, use a driver bolted to one of the T-slots. The driver can be an angle plate or any device that will prevent slippage in the chuck jaws. If multistep inside chuck jaws are used, the wheel can be clamped on the lower step to permit enough clearance above the table for the tool to turn the entire face without resetting (Fig. 5-10 and 5-11).

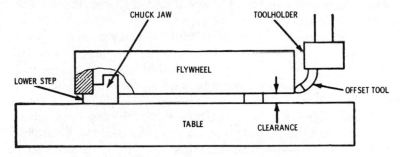

Fig. 5-10. Chucking a flywheel inside the rim. Also, machining by cylindrical turning on a boring mill.

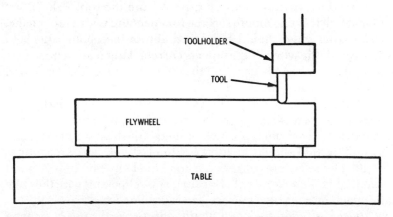

Fig. 5-11. Facing the side of the rim of a flywheel. This shows flat turning on a boring mill.

105

After the tool has faced the rim, it is moved over to face the hub. Then the hole for the shaft is bored, and the flywheel can be turned over so that the other side of the rim and hub can be faced. The chuck jaws can be removed and the finished side of the rim placed on the table to ensure parallelism for the last operation. After the flywheel has been turned over, it can be centered by means of a plug inserted in the hole of the table. The upper section of the plug should be of proper size to fit the hole in the flywheel, which is clamped against the spokes.

Conical turning—Usually, the saddle that carries the ram is arranged to swivel on a central stud by turning the angular adjustment gear which consists of a worm and segment. Provision is usually made for swiveling the ram to 45° on either side of the vertical; the amount of swiveling is indicated on a graduated scale. The worm and segment also act as a lock to prevent tipping and to permit turning the ram to any angle within its range. If a desired taper or conical surface is given in degrees, the ram can be set by means of a scale.

To turn a taper or a conical surface, swivel the ram to the given angle, and use the vertical feed. A setup for turning a gasoline engine flywheel having an inside conical surface for a cone clutch is shown in Fig. 5-12. The flywheel is held by a chuck and should be centered accurately, and the ram should be swiveled to the correct angle for the conical surface. The vertical feed should be set to feed downward, starting the cut from the top. Take a first cut that will leave a smooth surface for checking with a taper gage.

The taper gage should be placed across the center and held square with the work. If the taper is correct, light cannot be seen at any point between the gage and the conical surface. Take a second roughing cut, and check again with the taper gage. If the angle is incorrect, the angular adjustment setting must be changed.

After the correct setting has been obtained, take the first finishing cut, leaving about 0.010 inch of metal for the last finishing cut. A finishing tool should be used for the last cut, and precautions should be taken not to remove too much metal. It is a good practice to take several light finishing cuts, checking with the gage after each cut until the correct size is obtained.

Conical turning by combination of vertical and horizontal feeds—If the angle of a conical surface is greater than the angular

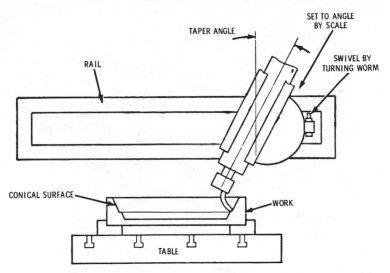

Fig. 5-12. Machining the flywheel of a gasoline engine. This shows conical turning on a boring mill.

range of the ram, another means of machining must be used. Simultaneous use of both the vertical and the horizontal feeds can be made to achieve the correct angle. Angular adjustment of the head is necessary; the proper setting involves a trigonometrical calculation.

According to F. D. Jones, the calculation can be made as follows:

1. Suppose a conical casting is to be turned to an angle of 30° (Fig. 5-13). The toolhead of the boring mill moves horizontally at 0.25 inch per revolution of the feed screw, and it has a vertical movement of 0.1875 (which is ³⁄₁₆) inch per turn of the upper feed shaft.

2. If the two feeds are used simultaneously, the tool will move a distance h of, perhaps, 8 inches while moving downward a distance v of 6 inches, thus turning the surface to an angle y. This angle is greater (as measured from a horizontal plane) than the angle required; but if the tool bar is swiveled to an angle x, the tool, as it moves downward, will also be advanced horizontally, in addition to the regular horizontal movement.

107

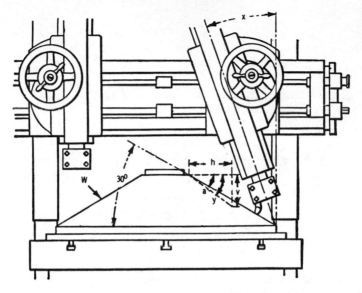

Fig. 5-13. Combination of the vertical and horizontal feed movements in turning a conical surface on the boring mill.

3. As a result, the angle y is diminished, and if the tool bar is set over the correct amount, the conical surface can be turned to an angle a of 30°. Then, the problem is to determine the angle x for turning to a given angle a.

The calculation for the angle x is explained in connection with the diagram in Fig. 5-14, which shows half the casting. Use a calculator to obtain accurate results.

1. The sine of the known angle a can be found in a table of natural sines. The sine of angle b, between the taper surface and the center line of the toolhead, can be determined by the equation:

$$\sin b = \frac{\sin a \times h}{v}$$

in which h represents the rate of the horizontal feed and v represents the rate of the vertical feed.

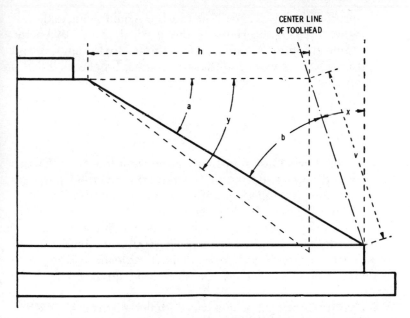

Fig. 5-14. Diagram showing a method of obtaining the angular position of the toolhead for turning a conical surface by combining the vertical and horizontal feed movements of the boring mill.

2. The angle corresponding to sine b can then be found in a table of sines. Thus, both angles b and a are known, and by subtracting the sum of these angles from 90°, the desired angle x can be obtained as follows:

a. sine of 30° = 0.500

b. sine of $b = \dfrac{0.500 \times 0.25}{0.1875} = 0.666$

c. angle $b = 41.759°$; and

angle $x = 90° - (30° + 41.759°) = 18.241°$

Therefore, to turn the casting to angle a in a boring mill having the horizontal and vertical feeds given, the toolhead should be set over 18.241° from the vertical.

3. If the required angle a were greater than angle y, obtained from the combined feeds with the tool bar in a vertical

109

position, the lower end of the tool bar would swing to the left rather than to the right of the vertical plane. When the required angle a exceeds angle y, the sum of the angles a and b is greater than 90°, so that the angle x for the toolhead is equal to $(a + b) =$ 90°.

SUMMARY

A vertical boring mill is virtually equivalent to a lathe that has been turned on end. Boring mills are used for various purposes such as drilling, reaming, chamfering, and counterboring. They can be classified as to the number of rams as either single ram or double ram, and they are available as either single- or double-station units. The size of a boring mill is designated by the size of the table. For example, a 72-inch mill has a table that is 72 inches in diameter. On engine lathes, the size designation is called the swing.

The vertical boring machine was designed on the principle that it is easier to lay a piece of work down than it is to hold it upright. Three general methods of fastening the work to the table are by bolts and clamps, by chucks, and by fixtures. Properly setting and fastening the work on the table is important in any boring operation.

The vertical boring mill can be used for turning as well as boring, especially for workpieces that require both turning and boring. The kinds of turning operations that can be performed on the boring mill are flat (facing), cylindrical, and conical.

REVIEW QUESTIONS

1. What makes the vertical boring mill different from the engine lathe?
2. What types of work can be performed on a vertical boring mill?
3. How is the size of a vertical boring mill determined?
4. What are the three methods of fastening the work to a boring mill table?

CHAPTER 6

The Horizontal Boring Machine

Although the vertical boring machine is adapted to boring and turning operations on work that has a diameter or width *greater* in proportion than its length; the horizontal boring machine is adapted to drilling, boring, and machining operations on work that has a diameter or width *smaller* in proportion to its length. The horizontal boring machine is very efficient for certain kinds of operations because nearly all the machining operations can be completed with just one setting on the machine (Fig. 6-1).

CLASSIFICATION

Horizontal boring machines can be classified according to their method of holding the work as either table- or floor-type

Courtesy Heald Machine Co.

Fig. 6-1. A horizontal boring machine. These machines can be used to machine large heavy work, such as cylinder blocks and large pump or compressor bodies.

machines. The boring head can be either of the stationary or vertically adjustable type. The table can be either adjustable vertically or nonadjustable vertically. Some special types of horizontal boring machines are used for machining railway motors, locomotive engine cylinders, etc.

Horizontal boring machines are rated as to size by the manufacturers, and each manufacturer can use his own method of rating. Generally, the size of the machine is given as the largest bar that the machine is designed to handle, but this may not be a true indication of the size of piece that the machine can handle for machining.

BASIC PARTS

As most horizontal boring machines are designed to machine large pieces of work, they usually have the various parts assembled on a heavy substantial bed. The head is mounted on a vertical column or post at one end of the bed.

Vertically Adjustable Head

The head post provides rigid support for the head in all operating positions. The entire head can be moved vertically on the ways of the column or head post (Fig. 6-2). As the vertical position of the head is changed, the outboard bearing also moves up or

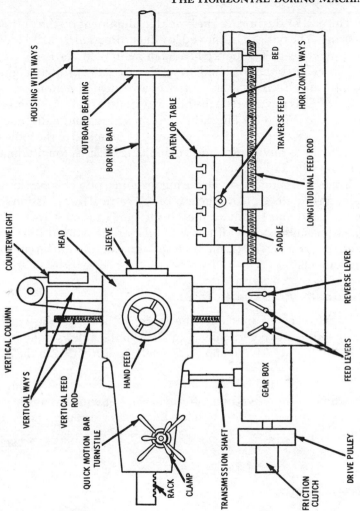

Fig. 6-2. Basic diagram of a horizontal boring machine having a head that is adjustable vertically.

down; the two parts are connected by shafts and gearing.

The head contains a sleeve in which a boring bar can move longitudinally. The bar carries cutters for boring operations, and either milling cutters or an auxiliary facing arm can be bolted to the end of the sleeve.

113

The end of the boring bar is held in alignment by the outboard bearing. The bar can be moved in either direction by hand feed or by power feed. The entire head can be moved vertically on the column ways either by hand, to set the bar at the proper height, or by power to feed a milling cutter in a vertical direction.

A saddle is mounted on the bed ways, and a table is mounted on the saddle (Fig. 6-3). The saddle can move longitudinally, and the table has a traverse feed, so that the work attached to the table can be moved to any position for machining, that is, longitudinally, transversely, and vertically.

In actual operation of the boring machine, power is transmitted through the speed change gear box and vertical transmission shaft to the head and finally to the boring bar, causing it rotate. The operator has a choice of many speeds for the work. Either hand feed or power feed can be used for all the different feed movements.

Vertically Adjustable Table

In this type of machine, the table, rather than the head, is designed to move vertically (Fig. 6-4). In the older machines, the spindle is cone driven, and back gears are provided in the same manner as on a lathe.

The head is fixed and cannot be moved vertically, but the spindle can be moved longitudinally in the cone by means of a pinion that meshes with a rack which traverses the spindle. This

Fig. 6-3. The table is mounted on the saddle, and the saddle is mounted on the bed ways of the horizontal boring machine.

114

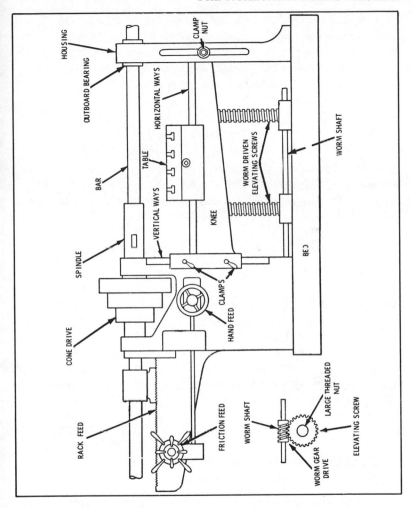

Fig. 6-4. Basic diagram of a horizontal boring machine having a table that is adjustable vertically. Detail of the worm gear drive is shown in the lower left.

movement can be effected by either a hand-feed or a power-feed mechanism.

As the spindle cannot be adjusted vertically, the work can be adjusted for height by raising or lowering the table. This can be accomplished by means of worm-driven elevating screws. The worms that move the elevating screws are attached to the worm shaft; hence each elevating screw is moved an equal amount. The knee, which carries the saddle and table assembly, traverses the vertical ways and is held firmly in any desired position by clamps at one end and clamp nuts at the other end.

OPERATIONS

In the following methods of machining, examples of drilling, boring, and facing operations are given to illustrate machining methods that are used on both the vertically adjustable spindle and the nonvertically adjustable spindle machines.

Drilling (without a Jig)

A length of cast iron pipe with flanges at both ends is used to illustrate drilling, facing, and boring operations that can be performed on the horizontal boring machine (Fig. 6-5). The dimensions, diameters of the bolt circles, and diameters of the holes are necessary for machining the piece.

The bolt circles on each flange must be laid out, marking the centers with a center punch. The work is mounted on V-blocks in the approximate position on the table of the machine, and

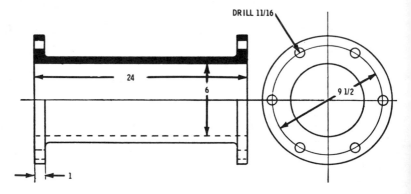

Fig. 6-5. Working drawing of a length of flanged pipe used to illustrate machining on a horizontal boring machine with a head vertically adjustable.

clamped lightly (Fig. 6-6). The position of the casting should be shifted so that two bolt-hole centers are at the same elevation above the table, that is, their scribed axis is horizontal, as determined by means of a surface gage. Then the casting must be aligned with the center line or spindle axis of the machine.

The boring bar should be passed through the casting and mounted in the spindle and outboard bearing; the height of the boring bar should be adjusted so that it is centered with the bore vertically. With the calipers, measure the distance from the bore to the boring bar at the horizontal axis, and compare with the distance at the other end of the bar. Shift the work until the bore is equidistant from the bar at both ends. Recheck the horizontal axis with the surface gage, and tighten the clamp bolts firmly. The horizontal axis should be trued, because two bolt holes can then be drilled without changing the setting of the vertical adjustment of the spindle.

Remove the boring bar, and place a drill of the proper size in the

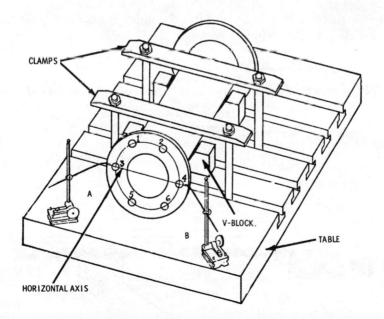

Fig. 6-6. Setup for fastening and aligning the flanged pipe on the table of the horizontal boring machine.

spindle. Use the cross feed screw to adjust the spindle vertically to hole No. 1 (see Fig. 6-6), and center the drill with the centerpunch mark. Start the machine and cut a trial circle. Back off the drill and check to be certain that the work is centered properly with the drill. Finish drilling the hole (using the spindle feed); then use the cross feed to shift the work, and drill hole No. 2. To drill holes No.3 and No. 4, lower the spindle to the correct height and drill as before. A third vertical adjustment can be made to drill holes No. 5 and No. 6.

Thus the six holes can be drilled with only three vertical settings of the spindle. If the axis of the two bolt holes is not exactly horizontal, a vertical setting for each hole is necessary.

Each hole should be spot-faced on the back of the flange so that the nuts can bear smoothly on the flange when they are tightened on the bolts. Remove the drill and insert a spot-facing bar into the spindle. Place the spot-facing bar in the last drilled hole, and insert a fly cutter having a cutting length twice the diameter of the hole (Fig. 6-7). Start the machine and, using slow feed, true or spot-face the surface of the flange around the hole. Similarly, repeat the operation for the other holes.

Drilling (with a Jig)

In production work where many pieces are to be machined, a jig

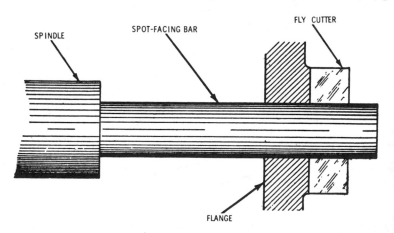

Fig. 6-7. Detail drawing of a spot-facing bar with fly cutter.

118

can save considerable expense because of the saving in time. Also, layout for each piece is avoided, and holes can be located and drilled with precision. The machined pieces are interchangeable, and do not have to be fitted when a jig is used to machine them.

To construct a jig for the flanged piece, the jig must be laid out from the given dimensions (see Fig. 6-5). When finished, the jig layout will appear as illustrated in Fig. 6-8. Scribe a horizontal axis *AB* onto the jig and through the centers of the two opposite bolt holes, and continue the line *BC* across the edge of the jig. Scribe the axis *AB* and *DE* on both flanges of the flanged pipe, and continue across the edges of the flange. The axis must be identical in its position on both flanges.

Mount the flanged pipe on the table of the horizontal boring machine, in the same manner as instructed for drilling without a jig, making certain that the axis *AB* is horizontal or parallel with the surface of the table. After the work has been aligned carefully and clamped securely to the table, place the jig over the flange that is to be bored, so that the scribed lines on the edge of the jig register with the scribed lines on the edge of the pipe flange; and clamp the jig to the pipe flange (Fig. 6-9). If the jig should shift while it is being clamped, it should be driven back into position with a lead hammer. In drilling, use the spindle feed. Perform the boring operation in the same manner as aforementioned for drilling without a jig.

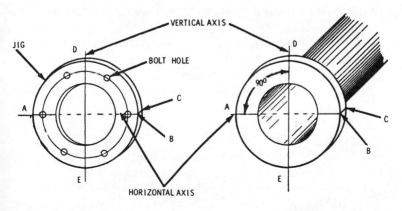

Fig. 6-8. Jig (left) and flange layout (right) for machining the flanged pipe.

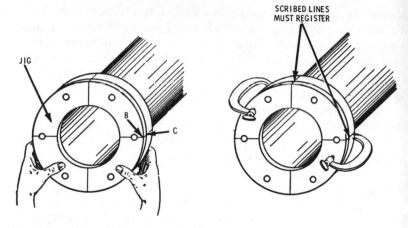

Fig. 6-9. Positioning the jig (left) and clamping the jig to the flange (right).

Boring

After all the bolt holes have been drilled, boring and facing operations can be illustrated, using the same flanged pipe (Fig. 6-10 and Fig. 6-11). Without unclamping the work (as it is already in alignment with the spindle axis), mount a boring bar in the machine and center it with respect to the pipe bore, using the vertical spindle feed and the horizontal table feed.

A variety of boring tools are available, such as a boring head and single- and double-end cutters. The selection of boring tools depends on the nature of the work, dimension of the bore, amount of metal to be removed, etc.

Move the table near the tailstock, as the table feed must be used. Mount the double-end cutter in the boring bar, and move the spindle with the hand feed until the cutter is placed at the beginning of the cut; then lock the spindle feed. The spindle feed should not be used for the boring operation, because the machine would bore out of true if the bar should become misaligned.

Take a roughing cut with a double-end cutter that is just the correct length to leave enough metal for the finishing cut (see Fig. 6-10). Use a single-end cutter for the finishing cut. If more than one finishing cut is necessary, the final cut should be very light (removing about 0.005 inch of metal). Note the shape of the finishing cutter in the detail drawing in Fig. 6-10.

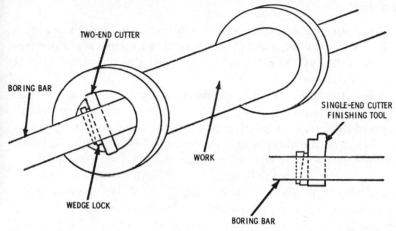

Fig. 6-10. A boring bar with a two-end cutter position for machining. Detail of the single-end cutter used for finishing is shown (lower right).

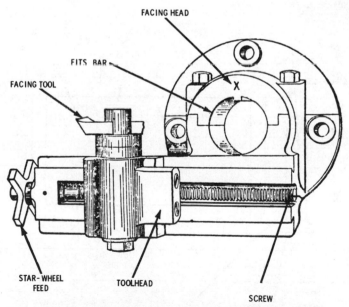

Fig. 6-11. The facing head is rotated by the bar. The star wheel contacts a stop pin clamped on the bed of the machine. As a point of the star wheel strikes the stop pin, rotating the wheel, the screw that actuates the facing tool is in turn rotated.

Facing

In all turning machines, the facing operation is flat turning, in which the work is traversed with a cutting tool in a *plane* perpendicular to the axis of rotation. A facing head, as shown in Fig. 6-11 can be used.

The setup for facing the pipe flange is shown in Fig. 6-12. When mounting the facing head, place it in position to feed inward toward the bar. Clamp a stop pin onto the table in such a position as to engage one of the spokes of the star wheel on each revolution of the bar, and feed the cutter across the surface to be faced.

Take a cut across the flange, deep enough to remove the scale, and true up the flange. Then bring the tool back to its original position; bring the work into position for the required depth of cut by means of the table feed, leaving enough metal for a light finishing cut. After finishing the first flange, transfer the facing head to the other end of the casting, and face the second flange in the same manner.

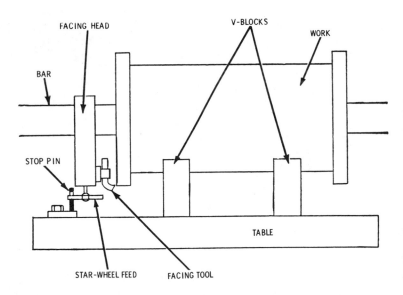

Fig. 6-12. Facing head setup for facing the pipe flange. Use a slow cutting speed (not more than 35 feet per minute).

SUMMARY

A vertical boring machine, as has been stated, is adapted to boring and turning work that has a diameter or width greater in proportion than its length. The horizontal boring machine is adapted to drilling, boring, and machining work that has a diameter or width smaller in proportion to its length.

Horizontal boring machines are generally classified according to their method of holding the work as either table or floor machines. Horizontal boring machines are rated as to size by the manufacturers, and each manufacturer can use his own method of rating. Generally, the size of the machine is given as the largest bore that the machine is designed to handle.

REVIEW QUESTIONS

1. What is the main difference between a vertical boring machine and a horizontal boring machine?
2. How are horizontal boring machines classified?
3. What determines the size of a horizontal boring machine?

CHAPTER 7

Lathes

The screw-cutting engine lathe is the oldest and most important of all the machine tools. Practically all of our modern machine tools have been developed from it.

DESIGN AND FUNCTIONS

Although lathes have changed greatly in both design and appearance, the fundamental principles of design, construction, and operation have remained virtually the same (Fig. 7-1). Therefore, a thorough understanding of the lathe and the work that can be performed on it enables a skilled worker to operate almost any lathe, regardless of its age or make.

The engine lathe produces cutting action by rotating the workpiece against the edge of the cutting tool. Lathe work is usually in the shape of a cylinder because the lathe produces a surface that is curved in one direction and is straight in one direction. The line of

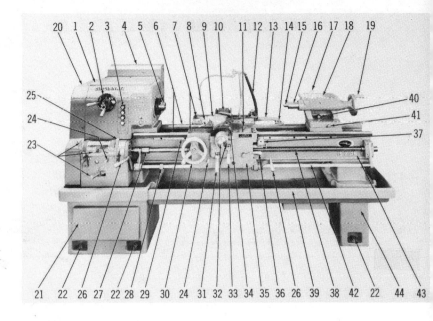

1. Headstock and tray
2. Spindle speed selector dial
3. Push button panel for start, stop, reverse, and coolant pump switch
4. Electrical compartment
5. Headstock spindle and spindle nose
6. Ways
7. Saddle (a component of the carriage assembly)
8. Compound rest dial (direct reading) and adjusting handle
9. Compound rest (a component of the carriage assembly
10. Back tool rest (part of cross slide)
11. Carriage system
12. Coolant system
13. Taper attachment bracket
14. Taper attachment clamp
15. Dead center in tailstock
16. Tailstock spindle
17. Tailstock spindle clamp
18. Tailstock and tray
19. Tailstock handwheel
20. Guard or cover for motor drive and gear train
21. Storage compartment
22. Leveling screw
23. Quick-change gear box and levers for settling lead and feed gears.

24. Oil shot system
25. Reversing lever for lead and feed
26. Headstock start and stop lever (two levers)
27. Bracket and clutch stop for longitudinal feed
28. Stop clamp for longitudinal feed
29. Rack for longitudinal feed
30. Handwheel for longitudinal feed
31. Longitudinal power feed engaging lever
32. Dial indicator (direct reading) and adjusting handle for cross slide
33. Cross slide power feed engaging lever
34. Cross slide (a component of the carriage assembly)
35. Apron (a component of the carriage assembly)
36. Half-nut engaging lever for threading
37. Thread dial for chasing dial
38. Lead screw and feed rod—feed is run off a keyway in the lead screw
39. Chip man
40. Clamp nut for tailstock
41. Adjusting screw for tailstock alignment or offset
42. Start and stop bar or control rod
43. Bed
44. Legs or base

Courtesy Cincinnati Milacron Co.

Fig. 7-1. Functional diagram of an engine lathe.

cut follows a curve (circumference of a circle), and the path of consecutive cuts produces a straight line to form the cylinder.

Size of Lathe

The maximum size of work that can be handled by the lathe is used to designate the size of the lathe, that is, the diameter and length of the work. Usually, the maximum diameter of the work is specified first, and the length is specified as the maximum distance between lathe centers (Fig. 7-2).

Manufacturers usually use the term *swing* to designate the size as the maximum diameter of the work that can be machined in the lathe. Thus, a 16-inch lathe indicates that the machine is designed to machine work up to 16 inches in diameter. Manufacturers also use the *length of the lathe bed*, rather than the distance between centers, to indicate lathe size.

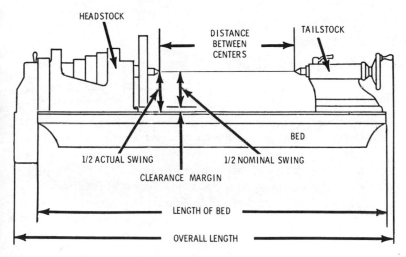

Fig. 7-2. Note the size designations of an engine lathe.

The swing of a lathe refers to the *nominal* swing and not to the actual swing of the lathe. The actual swing is the radial distance from the center axis to the bed and is always slightly greater than the nominal swing. This clearance margin can vary from ½ inch for a small lathe to 1½ inches for a large lathe to prevent any projection irregularities on the work coming in contact with the

127

lathe and resulting in damage. The rated swing of a lathe refers to the lathe bed and not to the lathe carriage. The swing over the carriage is necessarily less than the swing over the bed (Fig. 7-3).

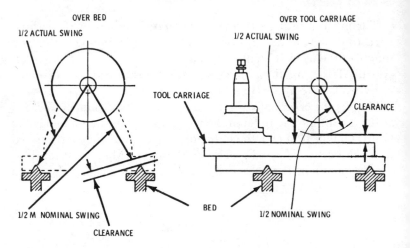

Fig. 7-3. The swing over the lathe bed (left), and over the carriage (right).

Types of Lathes

Engine lathes vary widely from the small bench lathe or *toolroom lathe* to the large *gap-bed lathes* and special-purpose lathes. Toolroom lathes are used to machine small parts. Larger and heavier lathes are used in high-production work. Examples of some of these high-production machines are turret lathes, multiple-tool lathes, and multiple-spindle lathes.

COMPONENT PARTS

A lathe is made up of many parts (see Fig. 7-1). The principal parts of a lathe are:

1. Bed.
2. Headstock.
3. Tailstock.
4. Carriage.

5. Feed mechanism.
6. Thread cutting mechanism.

Bed

The lathe bed (Fig. 7-4) is a stationary part that serves as a strong rigid foundation for a great many moving parts. Therefore, it must be scientifically designed and solidly constructed.

The *ways* of the lathe serve as a guide for the saddle of the carriage as it travels along the bed, guiding the cutting tool in a

Fig. 7-4. Lathe bed and ways of an engine lathe.

straight line. The V-ways (Fig. 7-5) are machined in the surface of the bed, and are precision finished to ensure proper alignment of all working parts mounted on the bed.

The two outer V-ways guide the lathe carriage. The inner V-ways and the flat way together provide a permanent seat for the headstock and a perfectly aligned seat for the tailstock in any position. A slight twist in the bed of the lathe can cause the machine to produce imperfect work. The lathe should be carefully leveled in both lengthwise and crosswise directions.

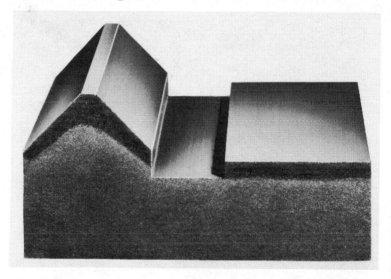

Courtesy Cincinnati Milacron Co.

Fig. 7-5. This is one of the V-ways and flat ways of a lathe bed.

Headstock

The headstock is mounted permanently on the bed of the engine lathe at the left-hand end of the machine. It is held in alignment by the ways of the bed and contains the gears which rotate the spindle and workpiece (Fig. 7-6).

Controlling headstock spindle speed—On a belt-driven *back-geared lathe*, the different steps of the cone pulley and the back gears are used to make changes in spindle speed. On the three-step pulley, three speeds can be obtained by driving directly from the

130

Courtesy Cincinnati Milacron Co.

Fig. 7-6. A lathe gearing mechanism enclosed in the headstock housing.

pulley to the spindle without the back gears in mesh. The smallest step gives the fastest speed. Three slower speeds can be obtained by driving the spindle through the back gears. Thus, three speed changes in direct drive and three speeds in back-gear drive permit a total of six changes of spindle speed.

The back gears can be meshed as follows?: (1) Stop the machine; (2) disengage the locking pin (the spindle will not move while the locking pin is disengaged, but the pulley will run freely); and (3) engage the back gears by pulling the handle forward to mesh the gears.

Various types of controls are used on the *geared-head* lathes (Fig. 7-7). Usually, selector levers are used, and the entire range of spindle speeds is obtained by reading directly from the dial.

Fig. 7-7. Spindle-speed selector on the headstock of a gear-head lathe. A "color match" dial is used to select the desired r/min.

Courtesy Cincinnati Milacron Co.

131

Spindle—The hollow spindle is built into the headstock with the *spindle nose* projecting from the housing of the headstock. The spindle is hollow throughout, and is tapered at the nose end to receive the live center. The spindle nose is usually threaded. Either a faceplate or a chuck can be turned on the threaded nose spindle to support and rotate the workpiece (Fig. 7-8).

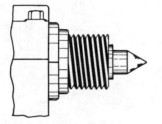

Fig. 7-8. A threaded spindle nose with a live center. A faceplate, a driver plate, or a chuck can be turned onto the threaded spindle nose to support the workpiece and to rotate it.

Drive plate—The driver plate turns onto the threaded spindle nose. It purpose is to drive the work mounted on the lathe centers by means of a *lathe dog*, which is clamped to the workpiece. The bent end of the dog engages a slot in the driver plate (Fig. 7-9).

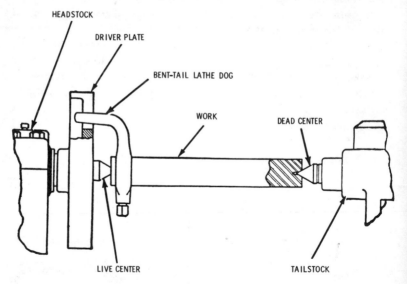

Fig. 7-9. Drive plate and bent-tail lathe dog used to drive work mounted between the lathe centers.

Tailstock

The tailstock assembly is movable on the bed ways, and carries the tailstock spindle (Fig. 7-10). The tailstock spindle has a standard Morse taper at the front end to receive a dead center. The tailstock handwheel is at the other end to give longitudinal movement when mounting the workpiece between centers. Reamers and taper-shank twist drills can be mounted in the tailstock spindle when required. A spindle binding lever clamps the spindle in any position in its travel. Clamp bolt nuts are used to clamp the tailstock assembly in any position of its travel on the ways of the bed. The dead center or any other tool mounted in the tailstock spindle can be removed by turning the tailstock handwheel counterclockwise.

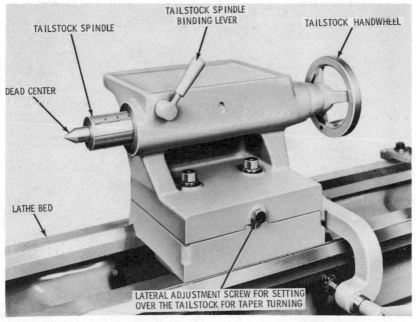

TAILSTOCK SPINDLE BINDING LEVER

TAILSTOCK SPINDLE

TAILSTOCK HANDWHEEL

DEAD CENTER

LATHE BED

LATERAL ADJUSTMENT SCREW FOR SETTING OVER THE TAILSTOCK FOR TAPER TURNING

Courtesy Cincinnati Milacron Co.

Fig. 7-10. Tailstock assembly.

Carriage

The carriage assembly is the entire unit that moves lengthwise along the ways between the headstock and the tailstock (Fig.

7-11). The carriage supports the cross slide, the compound rest, and the tool post. The two main parts of the carriage are the saddle (Fig. 7-12) and the apron.

Saddle—The saddle is an H-shaped casting that is machined to fit the outer ways of the lathe bed. The saddle can be moved along the ways either manually or by power, through the gearing mechanism in the apron. The apron and the cross slide are bolted to the saddle.

Cross slide—The cross slide is a casting that is mounted on and gibbed to the saddle (see Fig. 7-11). The cross-slide screw is located in the saddle and is connected to the cross slide. The cross slide can be moved either manually or by power across the saddle in a plane perpendicular to the longitudinal axis of the spindle.

Fig. 7-11. The carriage assembly including saddle, cross slide, compound rest, and apron.

The cross-feed micrometer dial is graduated in thousandths of an inch and is a very precise measuring instrument to be used when taking a cut on the work in the lathe. The operator must remember that when the dial is turned into the work one-thousandth of an inch, the diameter of the workpiece is reduced by two-thousandths of an inch because metal is removed from both sides of the continuously turning cylindrical work.

Compound rest—The compound rest is mounted on the cross slide (see Fig. 7-11). It consists of two main parts—the base and the slide. The base can be swiveled to any angle in a horizontal plane; the slide can be moved across the base by making hand adjustments with the micrometer dial, which is graduated in thousandths of an inch. The compound rest supports the cutting tool, and makes it possible to adjust the tool to various positions.

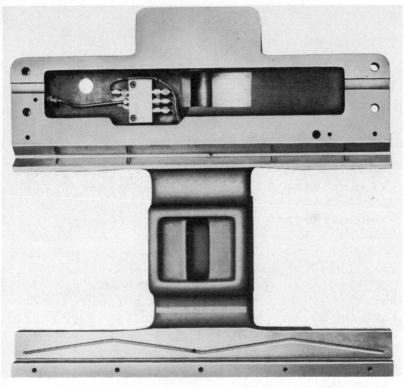

Courtesy Cincinnati Milacron Co.

Fig. 7-12. The saddle is the H-shaped part of the carriage assembly. It is machined to fit and is gibbed to the outer ways of the lathe bed.

The *tool post* assembly is mounted on the compound rest (Fig. 7-13). The rocker or wedge has a flat top, and is convex in shape on the bottom to fit into a concave ring or collar so that the cutting

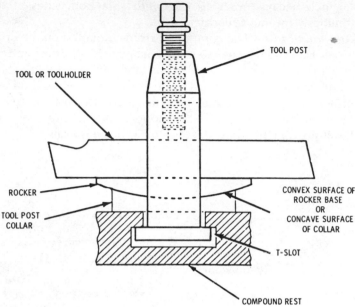

Fig. 7-13. Tool-post assembly mounted on the compound rest.

tool can be centered. A heavy-duty four-way turret tool post is shown in Fig. 7-14.

Apron—The apron is bolted to the front of the saddle. The apron houses the gears and controls for the carriage and the feed mechanism (Fig. 7-15 and Fig. 7-16). The carriage can be moved either manually or by engaging the power feeds.

The longitudinal feed handwheel, the longitudinal power-feed lever, the cross-slide power-feed lever, the half-nut lever, and the thread chasing dial are all located on the apron (see Fig. 7-15). The "Start-Stop" lever for the headstock spindle is usually located near the apron, convenient to the operator when he is standing in front of the apron.

Feed and Thread Cutting Mechanism

The same gears that move the carriage are involved in the feed and thread cutting mechanisms. These gears are used to transmit motion from the headstock spindle to the carriage.

Lead screw—Some lathes have two lead screws—one for turn-

Courtesy Cincinnati Milacron Co.

Fig. 7-14. A heavy-duty four-way turret tool post.

ing operations and one for thread cutting operations exclusively. The lead screw is very strong and has coarse, accurate threads (see Fig. 7-15).

Quick-change gear box—On most lathes, the quick-change gear box is located directly below the headstock on the front of the lathe bed (Fig. 7-17). A wide range of feeds and threads per inch may be selected by positioning the gears. The *index plate* is an index to the lever settings required to position the gears for the different feeds and numbers of threads per inch. The distance, in thousandths of an inch, that the carriage will move per revolution of the spindle is given in each block for the corresponding gear setting.

The *reversing lever* is used to reverse the direction of rotation of the screw for chasing right- or left-hand threads, and for reversing the direction of feed of the carriage assembly. Levers on the quick-change gear box should *never* be forced into position.

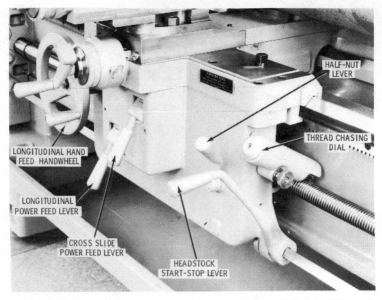

Fig. 7-15. Front view of the apron, which is bolted to the saddle, on the front of the lathe.

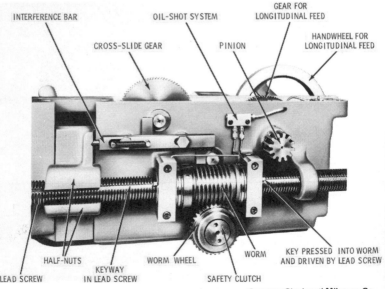

Fig. 7-16. Apron internal gearing (rear view).

TWO-POSITION LEVER

INDEX PLATE

REVERSING LEVER

INDEX PLUNGER

THREE-POSITION LEVER

HEADSTOCK SPINDLE START-STOP LEVER

Courtesy Cincinnati Milacron Co.

Fig. 7-17. Quick-change gear box. A wide range of feeds and threads per inch can be selected.

HOLDING AND DRIVING THE WORK

A number of work-holding devices are used to hold or support the workpiece securely for the different lathe operations. A number of different devices are also used as toolholders.

Chucks

Some workpieces must be held and rotated by chucks—depending on the size, shape, and operation to be performed. The chucks are attached to the headstock spindle of the lathe.

Great care should be exercised both in attaching and removing chucks from the headstock spindle. In attaching chucks to the *threaded spindle nose*, make certain that the spindle threads and the chuck threads are free from dirt and grit. The chuck turns onto the spindle in a clockwise direction—finish tightening the chuck with a quick spin by hand. Rotation of the workpiece against the cutting tool tightens the chuck on the spindle threads. A board or chuck block should be placed across the ways of the lathe to protect them when either mounting or removing the chuck.

Some lathes have a standard taper *key-drive spindle* (Fig. 7-18). In mounting, the lathe spindle should be turned so that the key is facing upward. Both the lathe spindle and the chuck taper should be free of grit, oil, and chips, especially along the spindle key. In positioning the chuck on the spindle, make sure the keys are in alignment before tightening the draw nut collar. Then, the draw nut collar should be tightened securely.

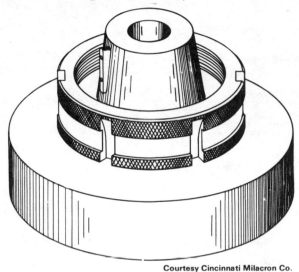

Courtesy Cincinnati Milacron Co.

Fig. 7-18. American Standard taper key-drive spindle.

Another type of spindle found on some lathes is the standard *camlock spindle* (Fig. 7-19). Here again, great care must be exercised in mounting and in removal to ensure accuracy and prevent damage.

Three-jaw, or universal, chuck—All three jaws of the universal chuck are moved at the same time in the chuck body. This chuck is usually used for either round or hexagonal stock. As the three jaws move simultaneously and are always the same distance from center, the workpiece is automatically centered without further adjustment (Fig. 7-20). However, some accuracy may be lost due to wear. To ensure accuracy in machining, a workpiece should

140

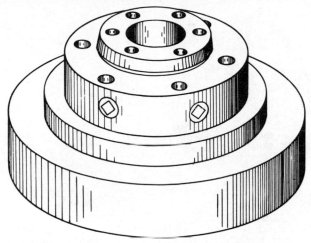

Fig. 7-19. American Standard camlock spindle.

Fig. 7-20.   Burned three-jaw, or
            universal, chuck.

never be removed or reversed until all operations on the work have been completed.

Four-jaw, or independent, chuck—The four jaws are adjusted independently in this type of chuck, and the jaws are reversible so that work of any shape can be clamped from either the inside or the outside of the jaws. The independent chuck is one of the most accurate means of chucking a workpiece (Fig. 7-21).

Mounting the work in a four-jaw chuck is largely a problem of centering. The concentric rings on the face of the chuck can be

141

Fig. 7-21. Burner four-jaw, or universal, chuck.

Courtesy Atlas Press Co.

used as a guide in centering the workpiece; the work should be clamped as closely centered as possible. Then test for trueness, marking the high spots with a piece of chalk resting against either the tool post or a tool mounted in the tool post (Fig. 7-22).

Brass strips or shim stock with a piece of emery cloth between the shim stock and the workpiece should be used on work having a finished surface to prevent the chuck jaws cutting into the surface finish. This also allows the work the be moved in the jaws for alignment.

Pointers on the use of chucks—Proper use and care of chucks is important. Some pointers that should be observed in handling and caring for these precision tools are:

1. Keep the chuck clean; do not oil excessively—a light film of oil on all working parts is ample.
2. Remove the live center and sleeve from the spindle before mounting the chuck.
3. Clean the face of the shoulder on the spindle nose and the back face of the chuck.
4. Place a few drops of oil on the threaded spindle nose.
5. Clean the threads in the chuck and on the lathe spindle with a clean brush before mounting a chuck on a threaded spindle nose.
6. When the chuck is close to the shoulder, give it a quick spin by hand to tighten it. A soft thud indicates a good firm seating against the shoulder. Forcing a chuck suddenly against the

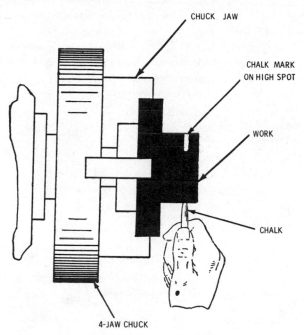

Fig. 7-22. Using a piece of chalk to center the workplace in a four-jaw, or independent, chuck.

shoulder strains the spindle and makes chuck removal difficult. The chuck is kept tight by rotation while the lathe is in operation.

7. Do not apply too much pressure while tightening the work in the chuck jaws. Excessive pressure may spring the work and affect the accuracy of the chuck.

8. Tighten the jaws around the more solid parts of the workpiece; always use the chuck wrench made for the chuck.

9. Turn the work as the jaws are tightened for a good fit.

10. Small-diameter work should not project more than four or five times the workpiece diameter from the chuck jaws—cuts should be short and light.

11. Heavy cutting pressures often cause the work to spring out and "ride the tool." Long work should be supported by the tailstock center, a steady rest, or a follower rest.

12. Do not force the chuck to carry work larger in diameter than

the diameter of the chuck body. Repeated overloading can damage the chuck.

13. If the jaws stick, tap lightly with a piece of wood and move them in and out to remove chips, this indicates that the chuck should be taken apart for a thorough cleaning. An old toothbrush is excellent for cleaning a chuck. Wash the parts in kerosene. Do not use excessive oil in reassembling the chuck; oil collects dust and chips, which clog the chuck mechanism.

14. Keep the chuck in a rack or bin when not in use. Dirt, dust, chips, and falling tools can cause much damage.

15. Place a board across the top of the ways to protect them from damage when either removing or mounting a chuck.

16. Never use an air hose to clean a chuck.

17. Never leave the chuck wrench in the work.

Draw collet chuck The *collet* is the holding or gripping part of the chuck. The assembly is comprised of a hollow draw-in spindle or bar which extends through the headstock spindle. The split collet is held in a tapered holding sleeve (Fig. 7-23).

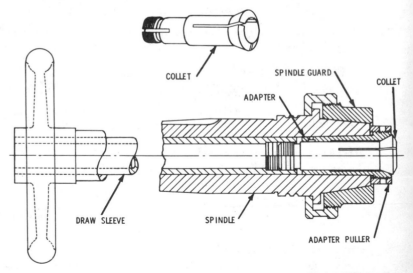

COLLET

SPINDLE GUARD

COLLET

ADAPTER

DRAW SLEEVE

SPINDLE

ADAPTER PULLER

Courtesy Cincinnati Milacron Co.

Fig. 7-23. A draw collet chuck as mounted on the headstock spindle.

144

A collet chuck is used for precision work in making small parts. The workpiece held in the collet should be not more than 0.001 inch smaller or 0.001 inch larger than the nominal size of the collet. A different collet should be used for each work diameter. If the work diameter is not within the above limits of the collet size, the accuracy and efficiency of the collet is impaired. To prevent distortion, a collet should never be closed unless a suitable plug or piece of work is inserted in the hole of the collet before it is closed.

Collets are available for various shapes of stock—round, square, etc. A new type of collet chuck and collet is shown in Fig. 7-24.

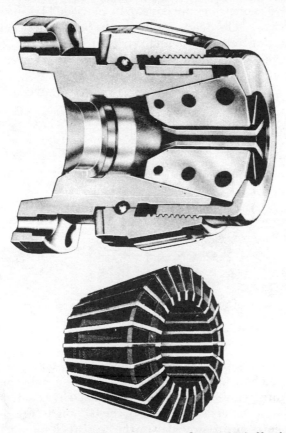

Fig. 7-24. Cutaway view of Jacobs collet chuck (top), and showing a rubberflex collet (bottom).

145

Faceplates—The workpiece can be mounted or held on a face-plate, which is a round metal plate that can be mounted on the headstock spindle (Fig. 7-25). The work is clamped or bolted to the elongated slots of the faceplate. The flat surface of the face-plate is perpendicular to the spindle.

Lathe centers—The *live center* is stationary in the main spindle of the headstock. It is called "live" because it turns with the workpiece. The *dead center* is placed in the tailstock spindle, and it does not rotate with the workpiece; hence the term "dead" center.

The live center does not have to be hardened because it rotates with the work. The dead center must be hardened because it does not rotate and must withstand the friction of the workpiece against it. Hardened centers are usually indicated by grooves cut in the circumference.

Fig. 7-25. Faceplate.

Courtesy Cincinnati Milacron Co.

Both centers usually have a standard Morse taper shank to fit in the tapered hole in the spindle, and a 60° cone end. The 60° angle may be turned on a lathe for the unhardened live centers; the dead or hardened centers are ground to the 60° angle (Fig. 7-26).

Live centers are removed from the headstock spindle by placing a knock-out bar through the hollow spindle to bump out the center. Care must be exercised to keep the center from striking the

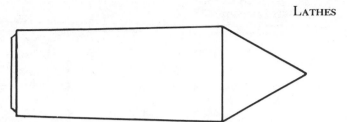

Fig. 7-26. A standard lathe center with 60° cone end.

ways of the lathe to avoid damage to either the center or the machine.

Some live centers are made for the machining of pipe and similar hollow-shaped work. The heads are constructed of hardened and ground steel to provide good resistance to abrasive wear. See Fig. 7-27. The bearings are protected by a two-piece seal, and accuracy is ± 0.0001 inch TIR. All pipe and bull heads come with a standard 60° included angle to match the chamfer on work for truer running.

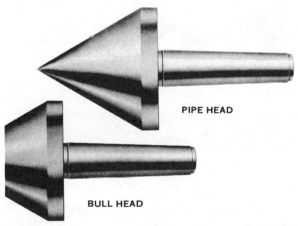

PIPE HEAD

BULL HEAD

Fig. 7-27. Pipe head and bull head live lathe centers.

Several types of centers are available for various purposes. A *female center* is shown in Fig. 7-28. The *ball center* (Fig. 7-29) is used when the "tailstock setover" method of turning tapers is used. If the standard live center is used in turning tapers by this method,

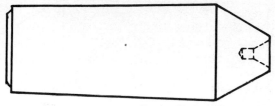

Fig. 7-28. A lathe female center.

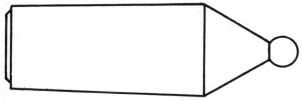

Fig. 7-29. Ball center.

proper bearing on both the center and the center hole may not be provided, resulting in damage to both the center and the workpiece.

The *half center* (Fig. 7-30) and the *crotch center* (Fig. 7-31) are used only as tailstock centers. The half center is used in machining small-diameter workpieces and in facing operations where it is

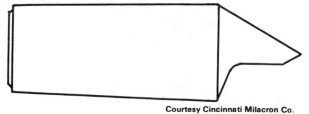

Fig. 7-30. Half center.

Fig. 7-31. Crotch center.

necessary to bring the facing tool near the center hole of the workpiece. The crotch center is used to hold a workpiece that is to be drilled at a location other than the ends. A revolving center (Fig. 7-32) rotates with the workpiece to eliminate the friction of the workpiece on the center. It is used in the tailstock of the lathe.

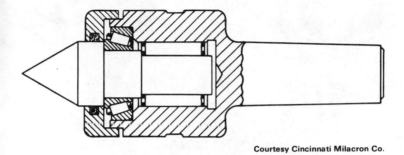

Courtesy Cincinnati Milacron Co.

Fig. 7-32. A revolving center used for the tailstock center.

Lathe dogs—The lathe dog is used with the drive plate as a simple means of causing the workpiece to rotate with the live center when the workpiece is mounted between centers (Fig. 7-33). The *straight dog* and the *bent-tail dog* are used for round work, and the *clamp dog* is used for work having a flat side.

The bent-tail dog and the clamp dog have a tail piece that is inserted in a slot of the driver plate. Shim stock should be used as a collar for the workpiece to prevent damage by the dog.

Mandrels—The standard lathe mandrel is a shaft or bar with 60° centers so that it may be mounted between centers. A workpiece is mounted on the mandrel for turning its outside surface true with the center hole. The mandrel is always rotated with a lathe dog; it is never placed in a chuck for turning the workpiece.

A *straight lathe mandrel* (Fig. 7-34) is designed to fit the entire length of the center hole in the workpiece. A *tapered mandrel* (Fig. 7-35) is used with parts that have a standard taper center hole. Workpieces having two internal diameters can be turned on a *double-diameter mandrel* (Fig. 7-36). The eccentric mandrel (Fig. 7-37) is used when an eccentricity is desired on the outside diameter of a workpiece.

149

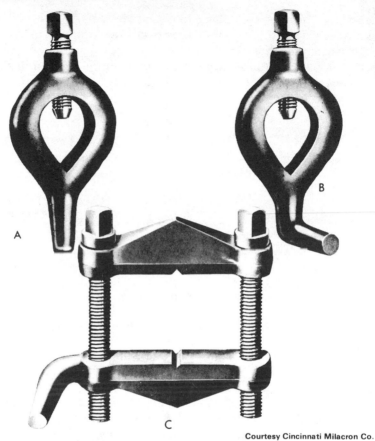

Courtesy Cincinnati Milacron Co.

Fig. 7-33. Lathe dogs: (A) straight dog; (B) bent-tail dog; (C) clamp dog.

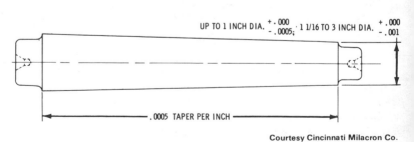

UP TO 1 INCH DIA. $^{+.000}_{-.0005}$; 1 1/16 TO 3 INCH DIA. $^{+.000}_{-.001}$

.0005 TAPER PER INCH

Courtesy Cincinnati Milacron Co.

Fig. 7-34. Straight lathe mandrel.

150

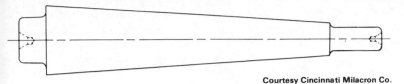

Courtesy Cincinnati Milacron Co.

Fig. 7-35. Tapered mandrel.

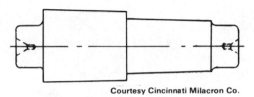

Courtesy Cincinnati Milacron Co.

Fig. 7-36. Double-diameter mandrel.

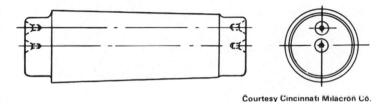

Courtesy Cincinnati Milacron Co.

Fig. 7-37. Eccentric mandrel.

SUMMARY

The lathe produces cutting action by rotating the workpiece against the edge of a cutting tool. Lathe work is usually in the shape of a cylinder because the lathe produces a surface that is curved in one direction and is straight in the other.

The maximum size of work that can be handled by the lathe is used to designate the size of the lathe. This is the diameter and length of the work. The manufacturer usually uses the term *swing* to designate the maximum diameter of the work that can be machined in the lathe.

Lathes vary widely from small bench models to the large gap-bed type and special-purpose machines. Toolroom lathes are usually used to machine small parts. The large lathes are used in high-production work. Some examples of these are turret lathes,

151

multiple-tool lathes, and multiple-spindle lathes. A lathe is made
up of many parts, such as the bed, headstock, tailstock, carriage,
feed mechanism, and thread-cutting mechanism.

REVIEW QUESTIONS

1. What determines the lathe size to use for a particular job?
2. Name the main components of a lathe.
3. Name three types of lathes generally used in high-production heavy-duty work.
4. What is the carriage section of a lathe?
5. Name the various types of chucks used on lathes.

CHAPTER 8

Lathe Operations

Proficiency in lathe operations involves more than just "turning" metal. The machinist should also have an understanding of necessary maintenance procedures and preliminary operations necessary to keep the lathe in good working condition.

Precision of any lathe, regardless of size, is greatly dependent upon the rigidity of the base under the lathe bed. Lighting should be adequate, and there should be enough space around the area for freedom in movement.

The engine lathe should be set level and remain level. If the lathe is not leveled, it will not set squarely on its legs, and the lathe bed will be twisted. If the headstock and V-ways are out of alignment, the workpiece will be tapered.

Wide metal shims and a sensitive level at least 12 inches in length are preferred for leveling a lathe. The bubble of the level should show a distinct movement when a 0.003-inch shim is placed under one end of the level. Both the headstock end and the tailstock end of the lathe should be leveled across the bed ways. Leveling should be repeated after the legs have been bolted down. Most machinists check the lathe for the level position regularly, and whenever the lathe is expected to be used for a long period of time for a particular job. The level position should also be checked before heavy work is attempted, and whenever the lathe is moved to a new shop location.

PREMACHINING OPERATIONS

Good finished work can be turned out on the lathe if the job is planned in advance. All parts are manufactured in a given operational sequence. Lathe operations are not difficult if the work is planned properly.

Cutting Speeds

Lathes are provided with a range of speeds that is ample to meet all conditions. Success in lathe work depends on the proper cutting speed. A cutting speed that is too slow wastes time and leaves a rough finish; a speed that is too high burns the cutting tool.

The cutting speed (tangential velocity) is the speed in feet per minute at which the surface of the work passes the cutting tool. The spool of thread analogy can be used to represent tangential speed as applied to lathe operations (Fig. 8-1). The familiar operation of pulling thread off a spool can be used to represent the speed of the revolving surface of the work in a lathe as it passes the cutting tool.

The cutting speed (feet per minute) at which the work passes the cutting edge of the tool is the chief factor in determining selection of the proper headstock spindle speed (r/min) of the lathe for a given material. Proper cutting speed lengthens tool life and results in a better finish. The recommended cutting speeds for the various materials and operations are given in Table 8-1.

Cutting speeds can be calculated by the following formulas:

$$\text{Spindle r/min} = \frac{\text{cutting speed (ft/min)} \times 12}{\text{circumference of work }(\pi D)}$$

$$= \frac{3.82 \times \text{cutting speed}}{\text{diameter of work}}$$

$$\text{Cutting speed} = \frac{\text{r/min} \times \text{diameter of work}}{3.82}$$

Table 8-1. Cutting Speeds (Feet per Minute) for High-Speed Steel Tools

C.M.M. 320,1117,1112,1018	Cast Iron—60% Steel	S.A.E. 3115, 3310, 4615, 8617	S.A.E. 81-B-45, 8650, 3150, L.G.T.S., H.G.T.S., H.S.S.	S.A.E. 3450, 4340	Bronze	Amco Bronze	
30	20	20	20	20	60	20	Tap
40	34	36	34	24	20	20	Ream
80	58	62	50	36	198	50	Drill Solid
90	58	62	54	38	134	54	Rough Bore
90	64	66	66	48	180	66	Finish Bore
100	70	66	62	42	180	62	Rough Turn (Memorize cut speed in dotted outline)
120	78	80	66	48	218	66	Finish Turn
100	70	72	66	48	180	66	Rough Face
80	78	70	66	48	148	66	Cut Off
120	100	108	90	68	218	98	Finish Face
30	28	30	28	20	40	28	Knurl (use.010" or .020" feed)
120	80	108	98	68	120	98	Center Drilling
40	20	25	20	20	45	20	Chasing or Threading

Note: Increase cutting speed approximately three times for carbide tools.

Courtesy Cincinnati Milacron Co.

Feed

Feed is the distance (in thousandths of an inch) that a tool advances into the work during one revolution of the headstock spindle. The maximum feed that is practical should always be used, considering the work, finish and accuracy required, and the rigidity of the work. Maximum efficient production is obtained only by familiarity with the work, the tools, and the lathe. The following formulas can be used to determine feed and time consumed for the operation.

$$F = \frac{L}{N \times T} \; ; \text{ and } T = \frac{L}{F \times N}$$

in which:

> F is equal to feed in thousandths of an inch,
> L is equal to length of cut,
> N is equal to r/min,
> T is equal to time in minutes.

Depth of Cut

To avoid lost time and wasted horsepower, the maximum amount of metal should be removed per cut per horsepower of the lathe. Cutting speed, feed, and depth of cut determine the amount of metal that is removed. As the operator becomes more familiar with the lathe and the work, he is more able to judge and use the maximum depth of cut.

Checking Alignment of Lathe Centers

If the lathe centers are not properly aligned, a tapered surface rather than a cylindrical surface will result. The lathe centers can be checked for alignment prior to turning by placing the dead center of the tailstock in close contact with the live center (Fig. 8-2). If the tailstock center does not line up, loosen the tailstock clamp bolt, and move the tailstock top in the proper direction by adjusting the tailstock setover screws to an approximate setting.

The final test for alignment of centers is illustrated in Fig. 8-3. Turn two 1.5-inch collars *A* and *B* on a piece of round stock that has

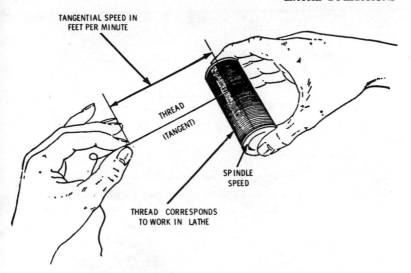

TANGENTIAL SPEED IN
FEET PER MINUTE

THREAD
(TANGENT)

SPINDLE
SPEED

THREAD CORRESPONDS
TO WORK IN LATHE

Fig. 8-1. The surface speed of the spool as the thread is unwound is comparable to the surface speed of the workpiece as the metal is removed by the cutting tool.

been centered. After a fine finishing cut, measure collar A. Then, without changing the caliper setting, measure collar B. Compare the diameters of the two collars. If the diameters are not the same, the centers are not in alignment; and the tailstock must be adjusted in the proper direction. Tailstock adjustment is illustrated in Fig. 8-4. The top of the tailstock can be set over by releasing one of the tailstock adjustment screws and tightening the opposite adjustment screw. The marks on the end of the tailstock top and bottom indicate the relative positions of the top and bottom (see Fig. 8-4). The alignment of these marks cannot be depended upon for fine, accurate work; the alignment test should be made to make certain that the lathe centers are aligned properly.

Centering the Workpiece

Center drilling is a very important premachining operation because the piece to be turned is supported on the lathe centers. The center holes are prepared by drilling and countersinking a hole in each end of the workpiece. The countersunk holes should have the same angle as the lathe centers that fit into them. The

157

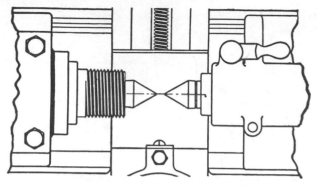

Fig. 8-2. Preliminary method of checking alignment of lathe centers.

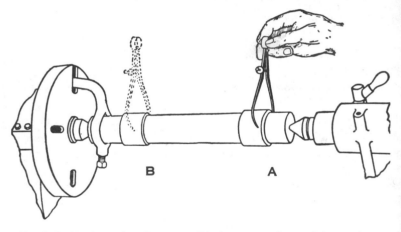

Fig. 8-3. Final test for alignment of lathe centers for straight turning.

standard angle is 60°, and the center hole should be deep enough that the point of the lathe center does not strike the bottom of the hole. A combination drill and countersink is used for this purpose. The center drill should penetrate in to the work $\frac{1}{64}$ (0.15625) inch past the 120° countersink (Fig. 8-5) for the lathe center to fit properly (Fig. 8-6).

In the center drilling operation, the stock is usually placed in the universal chuck. Not more than four or five times the diameter of the stock should be permitted to extend from the chuck. If the work is longer, a steady rest can be used for support—relieving the

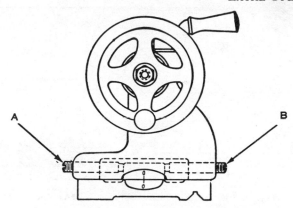

Fig. 8-4. Adjustment of the tailstock for alignment of the lathe centers by adjusting the setscrews (A and B).

chuck jaws of excessive stress. The combination drill and countersink is held in a drill chuck mounted in the tailstock spindle (Fig. 8-7).

BASIC OPERATIONS

The first and simplest exercise for the beginner in lathe operations is turning a piece of round stock to a given diameter. The work is mounted between lathe centers, and a lathe dog is used to drive the work.

Facing

After mounting the work between centers, the facing operation is performed. If an end is to be faced, the operation is more easily performed with the work mounted in the universal chuck. The facing tool is used to machine the ends of the work and to machine shoulders on the work.

The facing tool should be set with the point exactly on center (Fig. 8-8). The cutting edge of the facing tool should be set with the point of the tool slightly toward the work (Fig. 8-9).

Roughing cuts with the facing tool can be made either toward or away from the center; however, the finishing cuts must be made

159

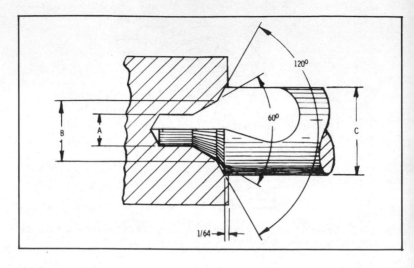

Stock Size	Drill No.	A = Drill Dia.	B = Bell Dia.	C = Body Dia.
1/8 – 1/4	11	3/64	0.100	1/8
1/4 – 7/16	12	1/16	0.150	3/16
3/8 – 5/8	13	3/32	0.200	1/4
1/2 – 1 1/4	14	7/64	0.250	5/16
1 1/4 – 1 3/4	15	5/32	0.350	7/16
1 3/8 – 2	16	3/16	0.400	1/2
1 2/2 – 2 1/2	17	7/32	0.500	5/8
Over 2 1/2	18	1/4	0.600	3/4

Fig. 8-5. Dimensions of holes that are center drilled with a combination drill and countersink.

Fig. 8-6. Center-drilled holes: (A) properly drilled; (B) incorrectly drilled (too deep); and (C) incorrectly drilled (not deep enough).

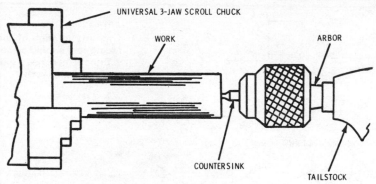

Fig. 8-7. Center drilling in round stock.

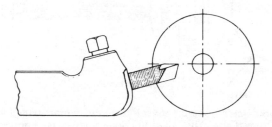

Fig. 8-8. The point of the facing tool should be set exactly on center.

by moving the point of the tool from the center to the outer edge of the cylinder. The point of the facing tool should be rounded slightly on an oilstone to give a smooth finish.

After one end is faced, the work should be reversed and the operation repeated, facing off just enough metal to bring the work to the specified length.

Turning

After the ends have been faced and center drilled, the work should be turned to the specified diameter.

Mounting on centers—In straight turning, a bent-tail dog is used to drive the work. It is important to select a lathe dog of correct size for the work and the slot of the driver plate. The tail of the dog should not bind at any point in the slot, nor should it touch the slot, except while driving the work—it should rock back and forth in the slot when mounted properly (Fig. 8-10).

161

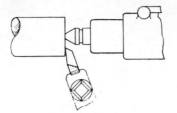

Fig. 8-9. The cutting edge of the facing tool should be set with the point turned slightly toward the work.

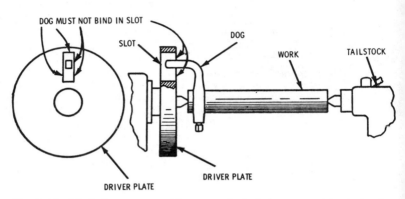

Fig. 8-10. Workpiece mounted between the lathe centers with a lathe dog of correct size. The tail of the dog must be free enough to rock back and forth in the slot in the driver plate.

The live center is used only for support of the work, and turns with the work; therefore, it requires no lubrication. However, the dead center must be lubricated as the dead center does not turn with the work, and a large quantity of heat can be generated. If the work is mounted too tightly or if lubrication is lacking, the tailstock center will burn. The work can be adjusted for play by means of the tailstock handwheel. The tailstock spindle should be clamped in position at the point where the lathe dog ceases to "chatter" when the lathe has been started.

Height of cutting tool—If the height of the cutting edge of a lathe tool coincides with the lathe axis, that is, the lathe centers, the tool is said to be on center; if higher or lower than the lathe axis, it is above or below center. The turning tool is usually set at a slight angle so that the tool will not dig into the work; if it should slip in the tool post, it will move away from the work (Fig. 8-11).

Theoretically, the correct height of a cutting tool is at a point

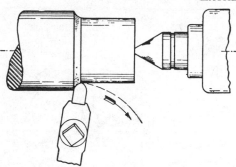

Fig. 8-11. Turning tool set at a slight angle so that it will not dig into the work if it should work loose in the tool post.

where a line through the front edge of the cutting tool coincides with a line tangent to the work; this corresponds to zero front clearance (Fig. 8-12). The theoretically correct height setting is ideal for a perfectly ground cutting tool; but as the tool becomes dull, the point rounds off slightly. Hence, the correct setting is actually slightly below the theoretically correct height setting.

The theoretically correct height setting decreases as the diameter of the work decreases (Fig. 8-13). As shown in Fig. 8-13, the tool

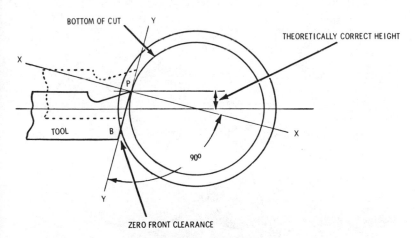

Fig. 8-12. Diagram showing the theoretically correct height of a cutting tool. If the front edge of the tool (PB) coincides with the line tangent to the work (YY), the tool is set at the correct theoretical height.

163

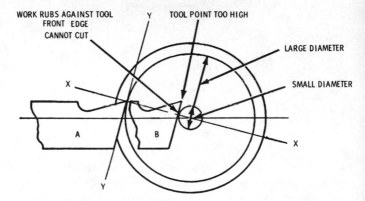

Fig. 8-13. A decrease in the theoretically correct height setting should be made with a decrease in the diameter of the work. Position A, which is correct for a larger diameter, is incorrect for a smaller diameter because the tool cannot cut at position B.

cannot cut at the smaller diameter because the original theoretically correct height setting for the large diameter is no longer correct as more metal is removed in turning the work.

In practical straight turning between centers, the cutting edge of the tool should be set at about 5° above center [or 3/64 (0.046875) inch per inch diameter of the work] as shown in Fig. 8-14. Taper

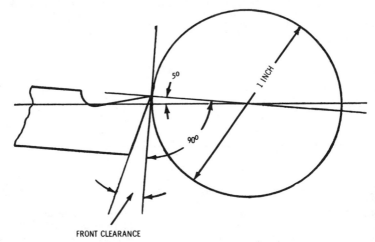

Fig. 8-14. Correct height setting of a cutting tool for straight turning.

164

turning, boring, and thread cutting also require the cutting tool to be set "on center." Materials that require the cutting tool to be set on center for turning are brass, copper, and other tenacious metals.

Roughing cuts—The roughing tool is used to remove most of the metal. This saves time and leaves only a small amount of metal to be removed by the finishing cut. Depending on the capacity of the lathe, only one roughing cut may be necessary. Of course, on a long, slender piece of work, the work may spring or break if too deep a cut is taken.

The roughing tool should project no more than necessary from the toolholder. Likewise, the toolholder should project only a short distance from the tool post.

After the roughing tool has been mounted in the tool post, the cross slide should be used to barely "touch up" the work. Then the cutting tool should be moved to the right-hand side, past the end of the rotating workpiece, and back into the workpiece—using the carriage handwheel. Just after the cutting tool enters the work, engage the power feed for the roughing cut from right to left. Disengage the power feed at about ⅛ (0.125) inch from the left-hand end of the cut, and feed the tool to the end of the cut (or shoulder) with the carriage handwheel. *Never* feed into a shoulder with the automatic feed because the carriage cannot be stopped at the correct point every time; thus, both the work and the tool could be ruined. Also, allow ample clearance between the end of the cut and the lathe dog.

After completing the cut with the carriage handwheel, return the carriage to the starting point at the right-hand side; adjust the cross slide micrometer dial for the depth of the next cut, and repeat the procedure until the workpiece is turned to the desired diameter. Leave approximately ¹⁄₃₂ inch (0.0312) for finishing the work smoothly to the specified size.

The cutting tool cannot cut to the extreme left-hand end of the work because of the lathe dog; the workpiece must be removed from the lathe, reversed, and placed in the lathe to machine the other end (Fig. 8-15). Shim stock should be used under the lathe dog to keep the setscrew from marring the turned end.

Finishing cuts—After the rough turning operation has been completed, the finishing tool can be mounted in the toolholder (Fig. 8-16). Care must be exercised in setting the cutting tool for

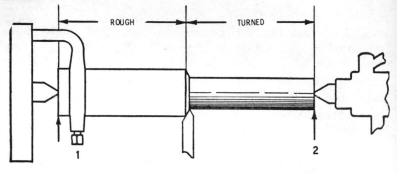

(A) First half of work is turned.

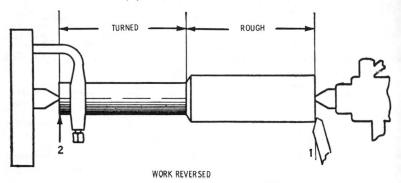

WORK REVERSED

(B) The lathe dog is removed and mounted onto the turned end.

Fig. 8-15. Reversing the workpiece to complete the turning operation between centers.

the finishing cut; if too much metal is removed, the diameter of the work will be too small, and the work will be ruined.

The depth of the finishing cut can be determined from the measurement taken of the rough turned work. It must be remembered that outside calipers are not precision measuring tools and are reliable only to within approximately 0.005 inch on work that is rough turned and not machined to finished size. Using the hand cross feed, the nose of the finishing tool should be brought in light contact with the rotating workpiece. Note the reading on the cross feed micrometer dial, and turn the carriage handwheel to the right until the tool is clear of the workpiece. Feed the cutting tool into the work with the hand cross feed—a distance equal to one-half

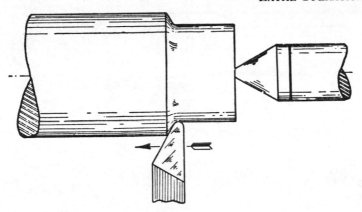

Fig. 8-16. Application of a finishing tool.

the difference between the rough-turned diameter and the specified finished size.

After selecting the proper cutting speed on the quick-change gear box, the automatic carriage feed is engaged for the finish cut. After moving a short distance past the end of the work, stop the lathe and measure the diameter with a micrometer caliper. If the work is over size, return the carriage to the right-hand side and adjust the cross-feed micrometer dial for the correct depth of cut. Proceed with the automatic carriage feed as close to the lathe dog as possible. Stop the lathe, remove the work, place the lathe dog on the opposite end of the workpiece, and complete the work to the specified size.

Several other turning tools can be used for machining work mounted between centers. Whenever possible, the right-hand facing, roughing, and finishing tools should be used because they move from right to left, which avoids placing extra friction or thrust against the dead center of the tailstock. The lathe headstock is mounted more sturdily, and the live center turns with the work; therefore, these parts are more capable of withstanding the heavy thrusts involved in turning.

In many instances, the left-hand tools (used for machining work from left to right) can be used more advantageously (Fig. 8-17). The round-nose tool can be used for turning in either direction. It is a convenient tool for reducing the diameter of a shaft in a central portion of the work (Fig. 8-18).

Cutoff tool—In many instances, stock is cut slightly longer than

167

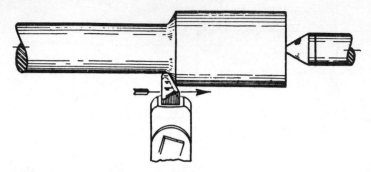

Fig. 8-17. Application of a left-hand tool.

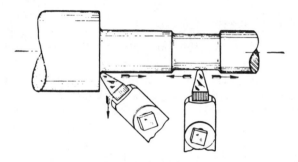

Fig. 8-18. Application of a round-nose tool. It is ground flat on top so that it can be used for turning in either direction of feed.

the finished dimension to allow for center drilling and facing to a finished length. Frequently, several different parts are machined on a single piece of stock and cut apart later. The cutoff tool is used for these purposes.

The cutoff tool should be mounted rigidly, with the carriage clamped to the lathe bed to keep the cutoff tool from moving either to the right or left. The tool should be fed into the work at right angles to the axis of the workpiece; and it should be set with the point of the tool exactly on center (Fig. 8-19). Cast iron requires no lubrication, but steel requires that the work be flooded with oil.

Work mounted on the faceplate—An engine lathe should have a large faceplate for mounting work that cannot be held conveniently in a chuck. The work is secured to the faceplate by clamp bolts that are anchored in the faceplate slots. This type of mounting is illustrated by the disk mounted directly to the faceplate as in Fig. 8-20. If the surface of the disk in contact with the faceplate is machined before mounting, a piece of paper should be placed between them to prevent slipping. However, if the rough casting is clamped to the faceplate, shims should be used to make the casting run true with the faceplate, that is, surface *A* should be parallel to surface *B*. Of course the work should be centered carefully with respect to the hub (see Fig. 8-20).

A round-nose tool should be used to face the hub surface *C*.

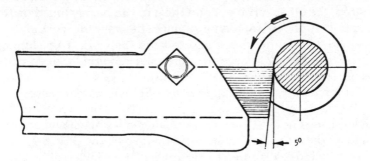

Fig. 8-19. Application of the cutoff tool. The tool point should be set exactly on center.

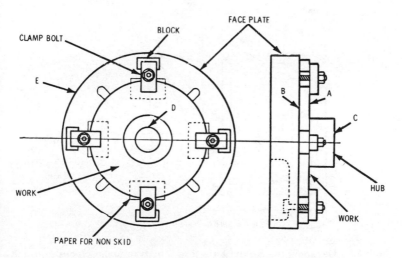

Fig. 8-20. Workpiece mounted on the faceplate for turning.

Then, the hole for the shaft D should be finished to the specific diameter with a boring tool.

If the hole for the shaft is larger than the central hole in the faceplate, parallel blocks should be used to set off the casting from the faceplate, so that the cutting tool will have clearance and will not contact the faceplate when finishing a cut. After this operation has been completed, the work can be removed from the faceplate and mounted on a mandrel, and the machining operation continued with the work mounted between centers. The rim should be turned to the finished diameter and the surface B faced. If desired, the other surface A and the hub can be machined on the mandrel. A "driver" should be mounted between the work and the faceplate to take the thrust and to keep the work from slipping when work having a large diameter and requiring a deep cut is being rotated.

Work mounted on the angle plate—Some irregular pieces cannot be machined by attaching them directly to the faceplate. The angle plate (Fig. 8-21) has two working sides machined at a 90° angle. In the setup for machining a flanged elbow (Fig. 8-22), the work is bolted or clamped to the angle plate, which is attached to the faceplate. Thus, if one finished flange is attached to the faceplate, the two flanged surfaces will be at 90°.

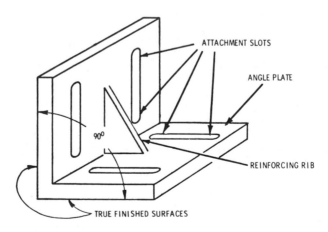

Fig. 8-21. The angle plate can be used to mount irregular pieces onto the faceplate for turning.

In actually performing the setup (see Fig. 8-22), the angle plate should be mounted on the faceplate so that the distance of the working surface of the angle plate from the lathe center is equal to the distance of one flange face from the axis of the other flange. This distance can be taken from the blueprint. If the elbow is centered laterally by adjusting its position on the angle plate, the facing cut will be concentric. It may be desirable to attach a counterweight, especially for light work, so that spindle speed can be increased.

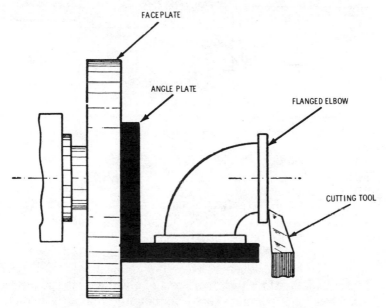

Fig. 8-22. Method of mounting an irregular workpiece, such as a flanged elbow, on an angle plate.

Work mounted on the carriage—Workpieces too large to mount on a faceplate, or in a chuck, can be anchored to the lathe carriage, and the cutting tool mounted between the lathe centers (Fig. 8-23). Large pieces—a pump cylinder, for example—can be bored by this method. After bolting the cylinder A to the saddle B of the lathe and mounting the cutter bar C between lathe centers, the cylinder can be bored by engaging the power feed which moves the cylinder toward the headstock.

171

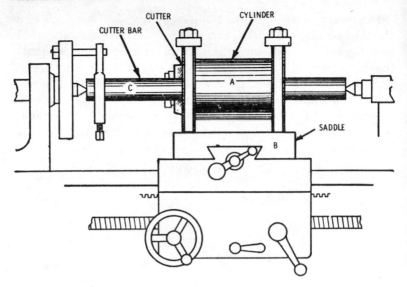

Fig. 8-23. Cylindrical workpiece mounted on the lathe carriage for boring.

In some instances, blocks or special fixtures bolted to the carriage can be used to hold the part to be bored. A cylindrical casting—a bushing, for example—can be mounted between wooden blocks bolted across the carriage. The blocks must be cut away to form a circular seat of correct radius, so that the work will be concentric with the boring bar.

The work to be bored should be mounted carefully and accurately on the lathe carriage. Circles should be laid off on each end of the cylinder to show the exact size and position of the bore when completed. The cylinder should be mounted temporarily on the carriage and the boring bar placed between the centers and through the cylinder (Fig. 8-24). With a pointed scriber wedged in the cutter slot on the boring bar, the boring bar should be rotated by hand and the cylinder adjusted in position until the scriber point is in contact with the layout circle on the cylinder at all points. This procedure must be repeated on the other end of the cylinder and the hold-down bolts must be tightened to fasten the work onto the carriage securely.

Drilling—The drilling operation can be performed very precisely on the lathe. The drill can be held stationary and the work

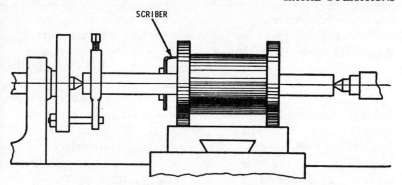

Fig. 8-24. Checking the alignment of a cylindrical workpiece mounted on the lathe carriage.

rotated; or the work can be held stationary and the drill rotated. The first method is more accurate than the second; the work is held in the chuck mounted on the headstock spindle, and the drill is held stationary in a drill chuck mounted in the tapered hole of the tailstock spindle. Then the drill is fed into the work with the tailstock handwheel.

As both the tailstock spindle and the headstock spindle have the same inside taper, the drill chuck arbor can be mounted in the headstock spindle for the second method of drilling; that is, the drill is rotated and the work is held stationary. The crotch center and the drill pad are two convenient accessories that can be mounted in the tailstock spindle for holding the work (Fig. 8-25).

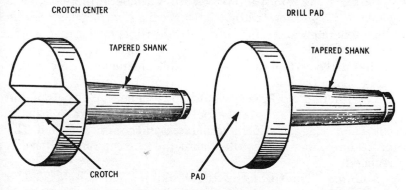

Fig. 8-25. Crotch center (left) and drill pad (right).

173

The crotch center automatically centers round work for cross drilling. The drill pad serves as a table for flat or square work. It is especially valuable for drilling large holes. The work should be held in the left hand and advanced against the drill by turning the tailstock handwheel with the right hand.

Reaming—The lathe can be used to ream holes to a very precise finished diameter. The holder for the reamer is held in the tailstock spindle. The reamer should be positioned at the end of the hole to be reamed. Then the headstock spindle should be started, and the reamer fed into the hole to the proper depth by means of the tailstock handwheel. Upon reaching the required depth, stop the lathe, and withdraw the reamer from the finished hole.

Holes are usually drilled a few thousandths under size for reaming to the finished size. A small drill of the finished size can be used up to $\frac{1}{16}$ (0.0625) inch in diameter; drills smaller than the finished size should be used for holes larger than $\frac{1}{16}$ (0.0625) inch in diameter. The following procedure is recommended for drilling and reaming holes: Use a center drill, a starting drill, a twist drill $\frac{1}{32}$ (0.03125) inch under the finished size, a twist drill $\frac{1}{64}$ (0.015625) inch under the finished size, and a reamer of the finished size.

For drilling, boring, and reaming holes up to $2\frac{1}{8}$ (2.125) inches in diameter, rough drill the hole approximately $\frac{1}{16}$ (0.0625) inch smaller than finished size; and ream to the finished size. For holes larger than $2\frac{1}{8}$ (2.125) inches, the work is usually bored to the finished size.

Boring—The removal of stock from a hole in the workpiece while it is being rotated by a lathe is called boring. This can be done accurately and efficiently with the proper boring tool, which must be mounted with the point exactly on center. The operator should make certain that there is ample clearance for the boring tool to enter the hole.

The rotating workpiece should be "touched up" by using the cross-feed hand lever, and the tool positioned for a light "cleanup" cut. Then, engage the longitudinal feed and complete the cut. The procedure must be repeated until the correct finished size is reached.

Knurling—The surface of the workpiece can be checked or roughened by rolling depressions into it by means of a knurling

174

tool (Fig. 8-26). In the knurling operation, the lathe dog is used to rotate the work. The knurling tool makes indentations in the surface in the form of straight lines or a diamond pattern (Fig. 8-27). Knurling provides a better hand grip on the metal and enriches the surface of the workpiece.

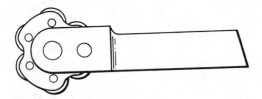

Fig. 8-26. A knurling tool.

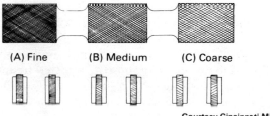

(A) Fine   (B) Medium   (C) Coarse

Courtesy Cincinnati Milacron Co.

Fig. 8-27. Knurled patterns (above) and knurling wheels (below) showing fine, medium, and coarse knurls.

The workpiece should be turned to the specified diameter before beginning the knurling operation. Then the section to be knurled should be turned to the diameter calculated as follows:

Number of teeth on circumference of work
$$= 66.8 \times (\text{outside diameter} - 0.017)$$

and:

Turned diameter $= 0.01496 \times$ number of teeth on the circumference of the workpiece

The knurling tool should be mounted in the tool post perpendicular to and on the center line of the workpiece. Feed the cross slide inward until the knurling tool contacts the section to be knurled. The longitudinal feed should be set at 0.010 inch or 0.020 inch on the quick-change gear box Then jog the lathe "on" and "off," while feeding the cross slide in, until the correct depth is attained. The appearance of the knurled surface determines the proper depth. Insufficient depth leaves flats on the diamond pattern; knurling too deeply causes nicks in the surface.

Several precautions should be observed in the knurling operation. The axles of the knurling wheels should be kept well lubricated to keep them turning freely; in setting the knurling tool, care should be taken to make certain that both wheels contact the surface evenly; and the knurling tool should be allowed to travel beyond the edge of the workpiece. Workpieces having a small diameter should be supported with a steady rest to prevent springing.

Turning an eccentric—An eccentric, by definition, is a circle or sphere not having the same center as another circle partly within or around it. A disk set off center on a shaft and revolving inside a strap attached to one end of a rod converts the circular motions of the shaft into a back-and-forth motion of the rod; thus, an eccentric can be used to obtain a reciprocating motion from a rotary motion. The distance between the center of the shaft and the center of the eccentric is the *eccentricity*; this is incorrectly called the *throw*, because the throw is twice the eccentricity.

In machining an eccentric with a hub, the centers of the eccentric should be laid out, the eccentric mounted on a faceplate and centered with respect to the center of the shaft through the hub, and a hole bored for the shaft. Then, the centers of the eccentric should be laid out on a mandrel of correct size for the hole in the eccentric (Fig. 8-28). The mandrel should be placed on V-blocks set on a surface gage and a line *AB* scribed through the mandrel center *O*. This is repeated on the other end of the mandrel. With the dividers set at the same length as the eccentricity, scribe an arc intersecting line *AB* and *C*, which is the center of the eccentric. Repeat the procedure at the other end of the mandrel. Then, the centers for the eccentric *C* can be drilled and countersunk.

After mounting the eccentric on the mandrel, place the mandrel in the lathe, with the lathe centers engaging the mandrel centers *O*, and machine the hub *D* and both sides *E*. Remove the mandrel and

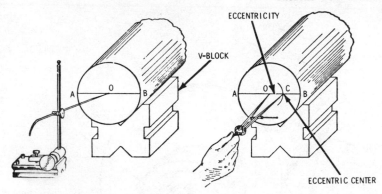

Fig. 8-28. A method of locating the centers of an eccentric on a mandrel.

reset in the lathe, with the centers of the eccentric C engaging the lathe centers; and machine the face F (Fig. 8-29).

Taper Turning

A taper is a cone-shaped surface—the diameter diminishing uniformly as the distance increases from the largest portion. Taper is measured as *taper (inches) per foot.* Thus, a tapered piece of work that has a two-inch diameter at one end and a one-inch diameter at the opposite end is said to have a taper equal to one inch per foot of length, or a *one-inch taper.*

A taper can be turned on a piece of work by means of one of the following methods:

1. Offset tailstock, thereby setting the lathe centers out of alignment.
2. Compound rest set at an angle.
3. Taper attachment which duplicates a set taper on workpiece.

Theoretically, the offset tailstock method is incorrect, but it can be used within limits (Fig. 8-30). The compound rest can be used to turn short tapers and bevels. The taper attachment, which will be discussed in a later chapter, is the approved method for turning work mounted between lathe centers.

Taper per foot—Length of the work and the difference in diameters of the two ends are the important factors in calculating taper. The formula for determining taper per foot is:

177

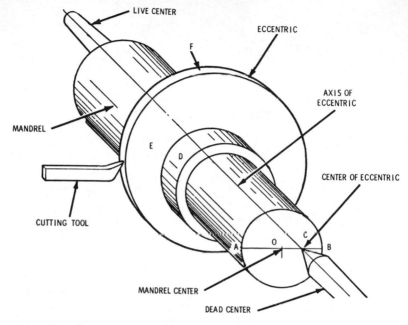

Fig. 8-29. Setup for machining an eccentric.

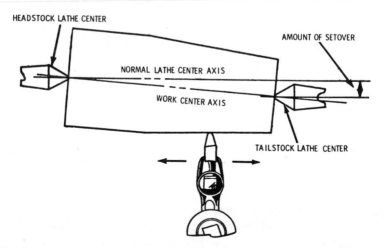

Fig. 8-30. Details of taper turning by the tailstock setover method. Note the direction of setover and direction of the taper.

$$\text{Taper per foot (inches)} = \frac{\text{major diameter} - \text{minor diameter} \times 12}{\text{length (inches)}}$$

Example: A piece of stock 10 inches in length is 3 inches in diameter at the large end, and 2½ inches in diameter at the small end. Calculate the taper per foot. Substituting in the formula:

$$\text{Taper per foot} = \frac{3 - 2.5}{10} \times 12$$

$$= \frac{0.5}{10} \times 12 = 0.6 \text{ inch}$$

Calculating tailstock setover—For a given taper, the tailstock must be set over a distance equal to one-half the total amount of taper for the entire length of the workpiece. If the tailstock setover is unchanged for workpieces of different lengths, the pieces will not have the same taper (Fig. 8-31).

The distance that the top of the tailstock must be offset depends on the desired taper per foot and the overall length of the work-piece. On a typical lathe, the upper part of the tailstock is locked in position by two headless setscrews (A and B in Fig. 8-32) located on the front and back of the base casting. To offset the tailstock, loosen the tailstock clamping nut, loosen the setscrew on the side toward which the tailstock is to be moved, and tighten the opposite setscrew. The index line C at the handwheel end of the tailstock can be used to indicate the amount of setover. The amount of

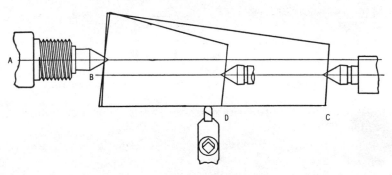

Fig. 8-31. Note the effect of the length of stock on taper per foot when the offset tailstock method is used.

179

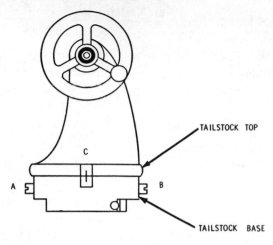

Fig. 8-32. Note the location of the setscrews for offsetting the tailstock to turn a taper.

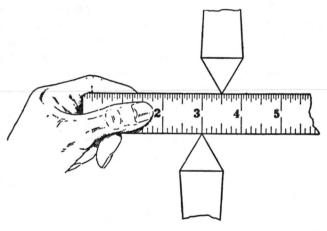

Fig. 8-33. A method of measuring the amount of tailstock offset.

tailstock setover can also be measured with a steel rule at the lathe centers (Fig. 8-33).

The following formulas can be used to calculate the amount of tailstock setover:

When the taper per foot is given,

180

$$\text{Setover} = \frac{\text{taper per foot}}{2} \times \frac{\text{length of taper in inches}}{12}$$

When the entire length of the piece is to be tapered, and the diameters at the ends of the tapers are given,

$$\text{Setover} = \frac{\text{major diameter} - \text{minor diameter}}{2}$$

When only a portion of the piece is to be tapered, and the diameters at the ends of the tapered portion are given, (see Note),

$$\text{Setover} = \frac{\text{total length of work}}{\text{length to be tapered}} \times \frac{\text{major diameter} - \text{minor diameter}}{2}$$

Note: A taper greater than one inch per foot on stock six inches or less in length is usually turned by tapering only a portion of a longer piece of stock, removing the waste portion after completing a taper. This avoids a small inaccuracy that results from misalignment of the angle of the lathe centers and the angle of the countersunk holes in each end of the stock.

Taper turning with the compound rest—This method is usually used for turning and boring short tapers and bevels—especially for bevel gear blanks and for die and pattern work. The compound rest should be set at the proper angle and the taper machined by turning the compound-rest feed screw by hand (Fig. 8-34). All angular or bevel turning should be checked with a gage as the operation progresses, as it is difficult to set the compound-rest swivel accurately for the desired taper.

Filing on the lathe—To produce a good finish or to bring the work down to size, filing may be resorted to; but it does not replace finishing by turning. If the work is within one or two thousandths of finished size and if specifications demand that the work be closer to finished size, the work can be filed while in the lathe.

When it is necessary to file on the lathe, the file should be held with the handle in the palm of one hand with the thumb on the top. The thumb of the other hand should be placed on top near the end of the file with the fingers curled over the end and gripping the file from underneath (Fig 8-35).

181

Only light pressure should be applied on the forward stroke, releasing the pressure on the back stroke without lifting the file from the work and taking slow even strokes. The file should not be held in one place on the work, but should be moved diagonally across the work, alternating the stroke from left to right and right to left.

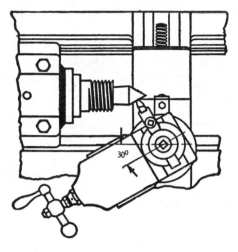

Fig. 8-34. The compound rest method of machining a 60° lathe center.

SUMMARY

Regardless of size, the precision of any lathe depends on the rigidity of the base under the lathe bed. There should be enough space around the area for freedom of movement and adequate lighting to perform all operations without eye strain.

Success in the work depends on the proper cutting speed. A cutting speed that is too slow can waste time and material, and a speed that is too high can burn the cutting tools. Lathes are provided with a range of speeds that is ample to meet all conditions. The cutting speed is the speed in feet per minute at which the surface of the work passes the cutting tool. The cutting speed at which the work passes the cutting edge of the tool is the chief

182

Courtesy Nicolson File Company

Fig. 8-35. Filing a workpiece mounted between centers on the lathe.

factor in selecting the proper headstock spindle speed for a given material.

Center holes are very important because the piece to be turned is supported on the lathe center with the help of these holes. The standard angle is 60°, and the center hole should be deep enough so that the point of the lathe center does not strike the bottom of the hole. After the ends have been faced and center drilled, the work should be turned to the specific diameter.

183

REVIEW QUESTIONS

1. Why are center holes drilled in pieces to be turned?
2. Why is the proper speed important in working with a lathe?
3. What is the standard angle for center holes?
4. What happens if the lathe centers are not properly aligned?
5. Why should a lathe be level?
6. What is meant by "feed"?
7. Why is the center drilled hole important on a piece of stock?
8. What is a facing operation on a lathe?
9. Why is the turning tool set at a slight angle?
10. What is the next step after completing the cut with the carriage handwheel?
11. How does a cutoff tool differ from a turning tool?
12. How is work held on a faceplate for turning?
13. What is the purpose of an angle plate?
14. How can drilling be done on a lathe?
15. How is reaming done on a lathe?

CHAPTER 9

Cutting Screw Threads on the Lathe

The screw thread is a very important mechanical device used for adjusting, fastening, and transmitting motion. Thread cutting in the engine lathe requires a thorough knowledge of thread cutting principles and procedures.

SCREW THREAD SYSTEMS

The threading operation actually involves cutting a helical groove of definite shape or angle, with a uniform advancement for each revolution, either on the surface of a round piece of material, or inside a cylindrical hole. Threads are either right-hand threads (advanced clockwise) or left-hand threads (advanced counterclockwise).

A number of thread systems are in use, which can result in confusion for a mechanic or machinist. Therefore, he should be familiar with the standard systems and their adaptations.

Screw Thread Forms

The most-used forms of screw threads have symmetrical sides inclined at equal angles with a vertical center line through the apex of the thread. The *Unified, Whitworth,* and *Acme* forms are examples of these forms. The *Sharp-V* was an early form of thread that is now used only occasionally. The symmetrical threads are easy to manufacture, and are widely used as general-purpose fasteners.

In addition, the so-called translation threads are used to move or translate machine parts against heavy loads. Thus, a stronger thread is required. Some of these translation threads that are widely used are the square, Acme, and buttress threads. The square thread is the most efficient of these threads; but it is the most difficult to cut because of its parallel sides, and it cannot be adjusted to compensate for wear. The Acme thread is only slightly weaker and less efficient, and it has none of the disadvantages of the square thread. The buttress thread is used to translate loads in one direction only, and it combines the ease of cutting and adjustment of the Acme thread with the efficiency and strength of the square thread.

Sharp-V thread—The top and bottom of the thread are theoretically sharp, but they are made slightly flat in actual practice. The sides form an angle of 60° with each other (Fig. 9-1).

American National screw thread forms—This form was the standard thread used in the United States for many years (Fig. 9-2). The thread form is practically the same as the American Standard for Unified threads now in use, except for certain modifications.

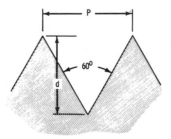

Fig. 9-1. Sharp-V thread form.

Courtesy *Machinery's Handbook*, The Industrial Press

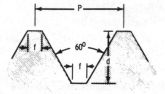

Fig. 9-2. American National thread form.

American Standard for Unified screw threads—In 1949, the American Standard Association approved the revision of the *American Standard Screw Threads for Bolts, Nuts, Machine Screws, and Threaded Parts* to cover the Unified screw threads on which the United States, Canada, and Great Britain reached agreement on nominal dimensions and size limits to obtain screw thread interchangeability among the three nations. The Unified threads are now the basic American standard for the fastening types of screw threads (Fig. 9-3). These threads have practically the same thread form and are mechanically interchangeable with American National threads of the same diameter and pitch. The principal differences in the two systems are the difference in

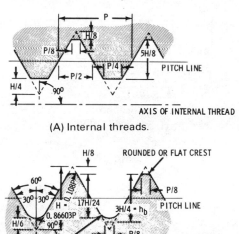

(A) Internal threads.

(B) External threads.

Fig. 9-3. Basic form of Unified screw threads.

amount of pitch-diameter tolerance on external and internal threads and the variation of tolerances with size.

In the American National system only the Class 1 external thread has an allowance, but in the Unified system an allowance is provided on both Classes 1A and 2A external threads. Also, the pitch-diameter tolerance of external and internal threads is equal in the American National system, whereas in the Unified system, the pitch-diameter tolerance of an internal thread is 30 percent greater than that of the external thread.

In the Unified system, the crest of the external thread can be either flat (preferred by American industry) or rounded (preferred by British industry). The crest is flat on internal threads. Otherwise, the Unified thread is practically the American National form long used in the United States.

Square thread—The Acme thread has largely replaced the square thread. Even though the square thread is stronger and more efficient, it is difficult to produce economically because of its perpendicular sides (Fig. 9-4).

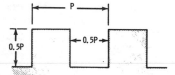

Fig. 9-4. Square thread form.

American Standard Acme screw threads—This standard provides three classes of *general-purpose* Acme threads for use in assemblies that require a rigidly fixed internal thread and have a lateral movement of the external thread limited by its bearing or bearings (Fig. 9-5). The standard also provides five classes of *centralizing* Acme threads.

American Standard Stub Acme screw threads—A Stub Acme screw thread is used for unusual applications where a coarse-pitch thread with a shallow depth is required.

American Standard Buttress screw threads—This form of thread is used where high stresses along the thread axis in one direction only are involved (Fig. 9-6). It is particularly applicable where tubular members are screwed together.

Dardelet thread—This is a patented self-locking thread de-

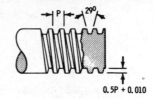

Fig. 9-5. Acme thread form.

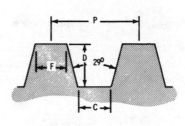

Fig. 9-6. Buttress thread form.

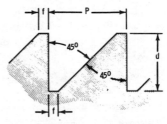

Courtesy *Machinery's Handbook*, The Industrial Press

signed to resist vibrations and to remain tight. The nut turns freely until it is seated tightly against a resisting surface. Then it shifts to the locking position due to a wedging action between the tapered crest of the thread of the nut and the tapered root of the thread of the bolt.

Whitworth Standard screw thread form—Since the Unified thread has been standardized, the British Whitworth form of parallel screw thread is expected to be used only for spare or replacement parts (Fig. 9-7).

International Metric thread system—This thread form was adopted at the International Congress for standardization of screw threads at Zurich, Switzerland, in 1898. It is similar to the American standard thread, except for the depth, which is greater.

American Standard for Unified miniature screw threads — These threads are intended for use as general-purpose fastening

189

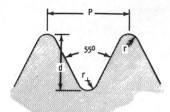

Fig. 9-7. Whitworth thread form.

Courtesy *Machinery's Handbook*, The Industrial Press

screws in watches, instruments, and miniature assemblies or mechanisms. Except for its basic height and depth of engagement, the threads are identical to the American and Unified basic thread form.

Screw Thread Series

Diameter-pitch combinations are grouped to form thread series, which are distinguished by the number of threads per inch as applied to a specific diameter. There are eleven standard series combinations.

Coarse-thread series (UNC)—The most commonly used series in the bulk production of bolts, screws, nuts, and other general applications is the coarse-thread series. It is used for threading materials having a low tensile strength, such as cast iron, mild steel, bronze, brass, aluminum, and plastics (Table 9-1).

Fine-thread series (UNF)—This series is suitable where the coarse series is not applicable for the production of bolts, screws, and nuts. The fine-thread series is used where the length of thread engagement is short and the wall thickness demands a fine pitch (Table 9-2).

Extra-fine thread series (UNEF)—This series can be used in most conditions applicable for fine threads. It is used where even finer threads are desirable, as for thin-walled tubes, couplings, nuts, or ferrules (Table 9-3).

SCREW THREAD TERMS

The mechanic or machinist should become familiar with the terms commonly used in connection with screw thread systems and thread cutting operations.

Table 9-1. Coarse-Thread Series, UNC and NC—Basic Dimensions (UNRC)

Sizes	Basic Major Diam.	Thds. per Inch	Basic Pitch Diam.	Minor Diameter		Lead Angle at Basic P.D.		Area of Minor Diam.	Tensile Stress Area
				Ext. Thds.	Int. Thds.	Deg.	Min.		
	Inches		Inches	Inches	Inches			Sq. In.	Sq. In.
1 (.073)*	0.0730	64	0.0629	0.0538	0.0561	4	31	0.00218	0.00263
2 (.086)	0.0860	56	0.0744	0.0641	0.0667	4	22	0.00310	0.00370
3 (.099)*	0.0990	48	0.0855	0.0734	0.0764	4	26	0.00406	0.00487
4 (.112)	0.1120	40	0.0958	0.0813	0.0849	4	45	0.00496	0.00604
5 (.125)	0.1250	40	0.1088	0.0943	0.0979	4	11	0.00672	0.00796
6 (.138)	0.1380	32	0.1177	0.0997	0.1042	4	50	0.00745	0.00909
8 (.164)	0.1640	32	0.1437	0.1257	0.1302	3	58	0.01196	0.0140
10 (.190)	0.1900	24	0.1629	0.1389	0.1449	4	39	0.01450	0.0175
12 (.216)*	0.2160	24	0.1889	0.1649	0.1709	4	1	0.0206	0.0242
1/4	0.2500	20	0.2175	0.1887	0.1959	4	11	0.0269	0.0318
5/16	0.3125	18	0.2764	0.2443	0.2524	3	40	0.0454	0.0524
3/8	0.3750	16	0.3344	0.2983	0.3073	3	24	0.0678	0.0775
7/16	0.4375	14	0.3911	0.3499	0.3602	3	20	0.0933	0.1063
1/2	0.5000	13	1.4500	0.4056	0.4167	3	7	0.1257	0.1419
9/16	0.5625	12	0.5084	0.4603	0.4723	2	59	0.162	0.182
5/8	0.6250	11	0.5660	0.5135	0.5266	2	56	0.202	0.226
3/4	0.7500	10	0.6850	0.6273	0.6417	2	40	0.302	0.334
7/8	0.8750	9	0.8028	0.7387	0.7547	2	31	0.419	0.462
1	1.0000	8	0.9188	0.8466	0.8647	2	29	0.551	0.606
1 1/8	1.1250	7	1.0322	0.9497	0.9704	2	31	0.693	0.763
1 1/4	1.2500	7	1.1572	1.0747	1.0954	2	15	0.890	0.969
1 3/8	1.3750	6	1.2667	1.1705	1.1946	2	24	1.054	1.155
1 1/2	1.5000	6	1.3917	1.2955	1.3196	2	11	1.294	1.405
1 3/4	1.7500	5	1.6201	1.5046	1.5335	2	15	1.74	1.90
2	2.0000	4 1/2	1.8557	1.7274	1.7594	2	11	2.30	2.50
2 1/4	2.2500	4 1/2	2.1057	1.9774	2.0094	1	55	3.02	3.25
2 1/2	2.5000	4	2.3376	2.1933	2.2294	1	57	3.72	4.00
2 3/4	2.7500	4	2.5876	2.4433	2.4794	1	46	4.62	4.93
3	3.0000	4	2.8376	2.6933	2.7294	1	36	5.62	5.97
3 1/4	3.2500	4	3.0876	2.9433	2.9794	1	29	6.72	7.10
3 1/2	3.5000	4	3.3376	3.1933	3.2294	1	22	7.92	8.33
3 3/4	3.7500	4	3.5876	3.4433	3.4794	1	16	9.21	9.66
4	4.0000	4	3.8376	3.6933	3.7294	1	11	10.61	11.08

Courtesy *Machinery's Handbook*, The Industrial Press

Allowance—A prescribed difference between the dimensions of mating parts. "Maximum allowance" is the difference between a minimum external and a maximum internal part. "Minimum allowance" is the difference between a maximum external and a minimum internal part. A "negative allowance" is the overlap, or

Table 9-2. Fine-Thread Series, UNF, UNRF, and NF— Basic Dimensions

Sizes	Basic Major Diam.	Thds. per Inch	Basic Pitch Diam.	Minor Diameter		Lead Angle at Basic P.D.		Area of Minor Diam.	Tensile Stress Area
				Ext. Thds.	Int. Thds.	Deg.	Min.		
	Inches		Inches	Inches	Inches			Sq. In.	Sq. In.
0 (.060)	0.0600	80	0.0519	0.0447	0.0465	4	23	0.00151	0.00180
1 (.073)*	0.0730	72	0.0640	0.0560	0.0580	3	57	0.00237	0.00278
2 (.086)	0.0860	64	0.0759	0.0668	0.0691	3	45	0.00339	0.00394
3 (.099)*	0.0990	56	0.0874	0.0771	0.0797	3	43	0.00451	0.00523
4 (.112)	0.1120	48	0.0985	0.0864	0.0894	3	51	0.00566	0.00661
5 (.125)	0.1250	44	0.1102	0.0971	0.1004	3	45	0.00716	0.00830
6 (.138)	0.1380	40	0.1218	0.1073	0.1109	3	44	0.00874	0.01015
8 (.164)	0.1640	36	0.1460	0.1299	0.1339	3	28	0.01285	0.01474
10 (.190)	0.1900	32	0.1697	0.1517	0.1562	3	21	0.0175	0.0200
12 (.216)	0.2160	28	0.1928	0.1722	0.1773	3	22	0.0226	0.0258
$\frac{1}{4}$	0.2500	28	0.2268	0.2062	0.2113	2	52	0.0326	0.0364
$\frac{5}{16}$	0.3125	24	0.2854	0.2614	0.2674	2	40	0.0524	0.0580
$\frac{3}{8}$	0.3750	24	0.3479	0.3239	0.3299	2	11	0.0809	0.0878
$\frac{7}{16}$	0.4375	20	0.4050	0.3762	0.3834	2	15	0.1090	0.1187
$\frac{1}{2}$	0.5000	20	0.4675	0.4387	0.4459	1	57	0.1486	0.1599
$\frac{9}{16}$	0.5625	18	0.5264	0.4943	0.5024	1	55	0.189	0.203
$\frac{5}{8}$	0.6250	18	0.5889	0.5568	0.5649	1	43	0.240	0.256
$\frac{3}{4}$	0.7500	16	0.7094	0.6733	0.6823	1	36	0.351	0.373
$\frac{7}{8}$	0.8750	14	0.8286	0.7874	0.7977	1	34	0.480	0.509
1	1.0000	12	0.9459	0.8978	0.9098	1	36	0.625	0.663
1 $\frac{1}{8}$	1.1250	12	1.0709	1.0228	1.0348	1	25	0.812	0.856
1 $\frac{1}{4}$	1.2500	12	1.1959	1.1478	1.1598	1	16	1.024	1.073
1 $\frac{3}{8}$	1.3750	12	1.3209	1.2728	1.2848	1	9	1.260	1.315
1 $\frac{1}{2}$	1.5000	12	1.4459	1.3978	1.4098	1	3	1.521	1.581

Courtesy *Machinery's Handbook*, The Industrial Press

interference, between the dimensions of mating parts. Allowance provides for variations in fit.

Angle of Thread—The included angle between the sides of the thread, measured in an axial plane.

Axis of Screw—The longitudinal central line through the screw from which all corresponding parts are equally distant.

Base of Thread—The bottom section of a thread; the largest section between two adjacent roots.

Basic—The theoretical or nominal standard size from which all variations are made.

Crest—The top surface joining the two sides of a thread.

Table 9-3. Extra-Fine-Thread Series, UNEF, UNREF and NEF—Basic Dimensions

Sizes	Basic Major Diam.	Thds. per Inch	Basic Pitch Diam.	Minor Diameter		Lead Angle at Basic P.D.		Area of Minor Diam.	Tensile Stress Area
				Ext. Thds.	Int. Thds.				
	Inches		Inches	Inches	Inches	Deg.	Min.	Sq. In.	Sq. In.
12(.216)*	0.3160	32	0.1957	0.1777	0.1822	2	55	0.0242	0.0270
¼	0.2500	32	0.2297	0.2117	0.2162	2	29	0.0344	0.0379
⁵/₁₆	0.3125	32	0.2922	0.2742	0.2787	1	57	0.0581	0.0625
³/₈	0.3750	32	0.3547	0.3367	0.3412	1	36	0.0878	0.0932
⁷/₁₆	0.4375	28	0.4143	0.3937	0.3988	1	34	0.1201	0.1274
½	0.5000	28	0.4768	0.4562	0.4613	1	22	0.162	0.170
⁹/₁₆	0.5625	24	0.5354	0.5114	0.5174	1	25	0.203	0.214
⁵/₈	0.6250	24	0.5979	0.5739	0.5799	1	16	0.256	0.268
¹¹/₁₆*	0.6875	24	0.6604	0.6364	0.6424	1	9	0.315	0.329
³/₄	0.7500	20	0.7175	0.6887	0.6959	1	16	0.369	0.386
¹³/₁₆*	0.8125	20	0.7800	0.7512	0.7584	1	10	0.439	0.458
⁷/₈	0.8750	20	0.8425	0.8137	0.8209	1	5	0.515	0.536
¹⁵/₁₆*	0.9375	20	0.9050	0.8762	0.8834	1	0	0.598	0.620
1	1.0000	20	0.9675	0.9387	0.9459	0	57	0.687	0.711
1¹/₁₆*	1.0625	18	1.0264	0.9943	1.0024	0	59	0.770	0.799
1⅛	1.1250	18	1.0889	1.0568	1.0649	0	56	0.871	0.901
1³/₁₆*	1.1875	18	1.1514	1.1193	1.1274	0	63	0.977	1.099
1¼	1.2500	18	1.2139	1.1818	1.1899	0	50	1.000	1.123
1⁵/₁₆*	1.3125	18	1.2764	1.2443	1.2524	0	48	1.208	1.244
1⅜	1.3750	18	1.3389	1.3068	1.3149	0	45	1.333	1.370
1⁷/₁₆*	1.4375	18	1.4014	1.3693	1.3774	0	43	1.464	1.503
1½	1.5000	18	1.4639	1.4318	1.4399	0	42	1.60	1.64
1⁹/₁₆*	1.5625	18	1.5264	1.4943	1.5024	0	40	1.74	1.79
1⅝	1.6250	18	1.5889	1.5568	1.5649	0	38	1.89	1.94
1¹¹/₁₆*	1.6875	18	1.6514	1.6193	1.6274	0	37	2.05	2.10

Courtesy *Machinery's Handbook*, The Industrial Press

Crest Clearance—Defined on a screw form as the space between the crest of a thread and the root of its component.

Depth of Engagement—The depth of a thread in contact, of two mating parts—measured radially.

Depth of Thread—The depth, in profile, is the distance between the top and the base of the thread (see Table 9-4).

External Thread—A thread on the outside of a member, as on a threaded plug.

Helix Angle—The angle made by the helix of the thread at the pitch diameter with a plane perpendicular to the axis.

193

Table 9-4. Double Depth of Threads

Threads per In. N	V Threads D D	Am. Nat. Form D D U.S. Std.	Whitworth Standard D D	Threads per In. N	V Threads D D	Am. Nat. Form D D U.S. Std.	Whitworth Standard D D
2	0.86650	.064950	0.64000	28	0.06185	0.04639	0.04571
2¼	0.77022	0.57733	0.56888	30	0.05773	0.04330	0.04266
2⅜	0.72960	0.45694	0.53894	32	0.05412	0.04059	0.04000
2½	0.69320	0.51960	0.51200	34	0.05097	0.03820	0.03764
2⅝	0.66015	0.49485	0.48761	36	0.04811	0.03608	0.03555
2¾	0.63019	0.47236	0.45545	38	0.04560	0.03418	0.03368
2⅞	0.60278	0.45182	0.44521	40	0.04330	0.03247	0.03200
3	0.57733	0.43300	0.42666	42	0.04126	0.03093	.03047
3¼	0.53323	0.39966	0.39384	44	0.03936	0.02952	0.02909
3½	0.49485	0.37114	0.35571	46	0.03767	0.02823	0.02782
4	0.43300	0.32475	0.32000	48	0.03608	0.02706	0.02666
4½	0.38438	0.28869	0.23444	50	0.03464	0.02598	0.02560
5	0.34660	0.25980	0.25600	52	0.03332	0.02498	0.02461
5½	0.31490	0.23618	0.23272	54	0.03209	0.02405	0.02370
6	0.28866	0.21650	0.21333	56	0.03093	0.02319	0.02285
7	0.24742	0.18557	0.13285	58	0.02987	0.02239	0.02206
8	0.21650	0.16237	0.15000	60	0.02887	0.02165	0.02133
9	0.19244	0.14433	0.14222	62	0.02795	0.02095	0.02064
10	0.17320	0.12990	0.12800	64	0.02706	0.02029	0.02000
11	0.15745	0.11809	0.11636	66	0.02625	0.01968	0.01939
11½	0.15069	0.11295	0.11121	68	0.02548	0.01910	0.01882
12	0.14433	0.10825	0.10666	70	0.02475	0.01855	0.01728
13	0.13323	0.09992	0.09846	72	0.02407	0.01804	0.01782
14	0.12357	0.09278	0.09142	74	0.02341	0.01752	0.01729
15	0.11555	0.08660	0.08533	76	0.02280	0.01714	0.01673
16	0.10825	0.08118	0.08000	78	0.02221	0.01665	0.01641
18	0.09622	0.07216	0.07111	80	0.02166	0.01623	0.01600
20	0.08660	0.06495	0.06400	82	0.02113	0.01584	0.01560
22	0.07872	0.05904	0.05818	84	0.02063	0.01546	0.01523
24	0.07216	0.05412	0.05333	86	0.02015	0.01510	0.01476
26	0.06661	0.04996	0.04923	88	0.01957	0.01476	0.01454
27	0.06418	0.04811	0.04740	90	0.01925	0.01443	0.01422

$$D D = \frac{1.733}{N} \text{ for V thread}$$

$$D D = \frac{1.299}{N} \text{ for Am. Nat. Form, U. S. Std.}$$

$$D D = \frac{1.28}{N} \text{ for Whitworth Standard}$$

Internal Thread—A thread on the inside of a member, as in a thread hole.

Lead—The distance a screw thread advances axially in one turn. On a single screw thread, lead and pitch are identical; on a double screw thread, lead is twice the pitch, etc.

Length of Engagement—The length of contact between two mating parts—measured axially.

Limits—the extreme dimensions prescribed to provide for variations in fit and workmanship.

Major Diameter—The largest diameter of the thread of a screw or nut. The term replaces "outside diameter" as applied to the thread of a screw and "full diameter" as applied to the thread of a nut.

Minor Diameter—The smallest diameter of the thread of a screw or nut. The term replaces "core diameter" as applied to the thread of a screw and "inside diameter" as applied to the thread of a nut.

Neutral Space—The space between mating parts that must not be encroached upon.

Number of Threads—The number of threads in one inch of length.

Pitch—The distance from a point on a screw thread to a corresponding point on an adjacent thread, measured parallel to the axis.

Pitch Diameter—On a straight screw thread, pitch diameter is the diameter of an imaginary cylinder, the surface of which would pass through the threads at such points as to make equal the width of the threads and the width of the spaces cut by the surface of the cylinder.

Root—The bottom surface joining the sides of adjacent threads.

Screw Thread—A ridge of a desired profile generated in the form of a helix either on the inside or on the outside surface of a cylinder or cone.

Side of Thread—The surface of the thread that connects the crest with the root.

Tolerance—The amount of variation permitted in the size of a part. It is the difference between the limits or maximum and minimum dimensions of a given part. Tolerance may be expressed as plus, minus, or as both plus and minus. Total tolerance is the sum of a plus and minus tolerance. Net tolerance is the total tolerance reduced by gage manufacturing tolerances and wear limits of the part.

CHANGE-GEAR CALCULATIONS

In the absence of an index chart showing gear combinations for various threads, it is necessary to calculate the proper gears to use for cutting threads. On lathes equipped with tumbler reverse gears, it must first be determined whether the lathe is *even geared* or *odd geared*.

On an even-geared lathe, the stud gear revolves at the same speed as the spindle gear, that is, the two gears are equal in size. If the stud gear revolves at any other speed, the lathe is an odd-geared lathe.

Change Gears

Change gears are either *simple* or *compound* in form. In simple gearing, an idler gear is used to transmit motion from the stud gear to the lead screw gear (Fig. 9-8). In compound gearing, the idler is combined with another gear of different size; and as the two gears are keyed together, the assembly no longer functions as an idler,

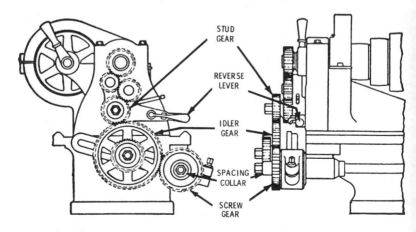

Fig. 9-8. Simple gearing for cutting screw threads on a change-gear lathe.

but as a first-stage ratio reduction between the stud gear and the lead screw gear (Fig. 9-9). This "compound gear assembly" is necessary when a large ratio between the stud gear and lead screw gear is required to cut extra-fine threads.

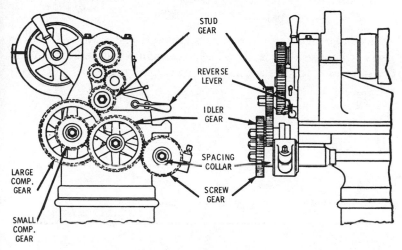

Fig. 9-9. Compound gearing for cutting screw threads on a change-gear lathe.

Simple gearing—The ratio between the number of teeth on the stud gear and the lead screw gear must be determined. The ratio depends on the pitch of the lead screw and the number of teeth to be cut. Expressed as a formula:

$$\text{Change-gear ratio} = \frac{\text{threads per inch on lead screw}}{\text{number of threads to be cut}}$$

Example: Determine the size of the stud gear and the lead screw gear required to cut 12 threads per inch in a lathe having a lead screw with 8 threads per inch.

Substituting in the formula:

$$\text{Gear ratio} = \frac{8}{12}$$

To cut 12 threads per inch, the spindle (or stud gear on an even-geared lathe) must make 12 revolutions to 8 revolutions of the lead screw. Thus, if gears of 12 teeth are available, they would cut the required thread. As these gears are impossible, multiply each by a common number to obtain the desired gears available in the change-gear set as follows:

1. Stud gear = 8 × 3 = 24 teeth
2. Lead screw gear = 12 × 3 = 36 teeth

As change gears of various sizes are used, the distance between gears will vary. Therefore, an idler gear must be used to transmit the motion (Fig. 9-10). The idler gear does not change the gear ratio.

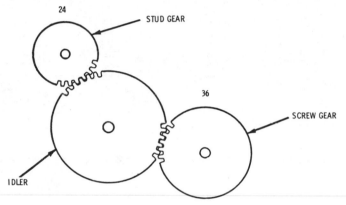

Fig. 9-10. Diagram of simple change gearing.

Compound gearing—When the gears are arranged in a train, they are said to be compounded. As in simple gearing, the gear ratio must be determined between the stud gear and the lead screw gear. To illustrate the necessity for compound gearing, suppose that it is desired to cut 80 threads per inch in a lathe having a lead screw with 8 threads per inch.

Substituting in the formula:

$$\text{Gear ratio} = \frac{8}{80}$$

To cut 80 threads per inch, the spindle (or stud gear in an even-geared lathe) must make 80 revolutions to 8 revolutions of the lead screw. It is impossible to use gears with 8 and 80 teeth to cut the required thread. Multiply the number of teeth by a common number to obtain gears within the range of the change-gear set. As a 16-tooth gear is usually the lowest number of teeth furnished in the set, multiply by 2 as follows:

1. Stud gear = 8 × 2 = 16 teeth
2. Lead screw gear = 80 × 2 = 160 teeth

As a 160-tooth gear is entirely out of range most equipment, and the required diameter would probably be too large to mesh with the stud gear, compound gearing is necessary so that smaller gears can be used.

Example: Determine the compound gears necessary to cut 80 threads per inch on a lathe having a lead screw with 8 threads per inch. The following procedure can be used to determine the required gears:

1. The ratio between the stud gear and the lead screw gear is as follows:

$$\text{Total ratio} = \frac{8}{80}$$

2. As this is a 10 to 1 ratio, the stud gear must make 10 revolutions to 1 revolution of the lead screw. Factor the total ratio as follows:

$$\frac{8}{80} = \frac{2 \times 4}{8 \times 10}$$

3. Multiply each term by a common number (8, for example,) and obtain the required gears as:

$$\frac{16}{64} = \frac{32}{80}$$

4. Place these gears in the following order: 16, 64, 32, and 80. Use the 16-tooth gear as the stud gear; use the 64- and 32-tooth gears for the compound gear assembly; and place the 80-tooth gear on the lead screw (Fig. 9-11). This gear setup will provide a 10 to 1 ratio or speed reduction between the lathe spindle and the lead screw.

Compound gearing actually consists of two ratios whose product is equivalent to the total ratio as follows:

$$\text{First ratio} = \frac{16}{64} = \frac{1}{4}$$

$$\text{Second ratio} = \frac{32}{80} = \frac{1}{2.5}$$

and:

$$\text{Total ratio} = \text{first ratio} \times \text{second ratio}$$

$$= \frac{1}{4} \times \frac{1}{2.5} = \frac{1}{10}$$

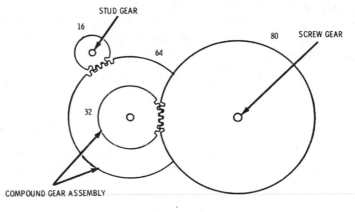

Fig. 9-11. Diagram of compound change gearing.

When compound gearing is used, the ratio of the compound gears is usually 2 to 1, so that the threads are twice the number per inch as when simple gearing is used. Therefore, the following procedure can be used:

1. Calculate as in simple gearing:

$$\text{Gear ratio} = \frac{8 \text{ (stud gear)}}{40 \text{ (lead screw gear)}}$$

2. Doubling these figures results in 16 teeth for the stud gear and 80 teeth for the lead screw gear respectively. Selecting, for example, 18- and 36-tooth gears for the compound assembly, the setup would be as follows: A 16-tooth gear for the stud gear; 36- and 18-tooth gears for the compound assembly; and

an 80-tooth gear for the lead screw (Fig. 9-12). Thus, the following ratios are:

$$\text{Stud gear to lead screw gear ratio} = \frac{16}{80} = \frac{1}{5}$$

$$\text{Compound assembly ratio} = \frac{18}{36} = \frac{1}{2}$$

and:

$$\text{Total ratio} = \frac{1}{5} \times \frac{1}{2} = \frac{1}{10}$$

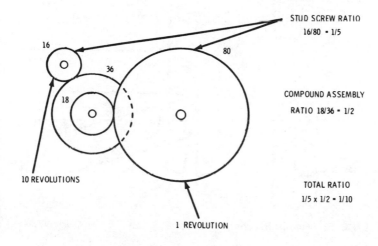

Fig. 9-12. Diagram showing use of total ratio in a compound change-gear setup.

Quick-Change Gear Lathes

The quick-change gear box permits the operator to obtain the various pitches of threads without using loose gears. All lathes equipped with quick-change gear boxes have index plates or charts for setting up the lathe to cut various threads. It is necessary only to arrange the levers on the gear box for the various threads per inch, as indicated on the index plate.

THREAD CUTTING TOOLS

The shape of the cutting tool depends on the type of thread to be cut. The tool must be ground and set accurately or the correct thread form will not be tolerated. Various gages are available that can be used as guides in grinding.

The shape of the cutting tool for cutting a sharp-v 60° thread is shown in Fig. 9-13. Sometimes the point is not ground off, and the thread is cut with the sharp-v bottom (obsolete); this should never be done when maximum strength is desired.

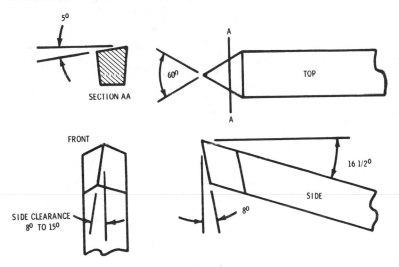

Fig. 9-13. Tool for cutting sharp-v threads.

The Acme screw thread form is often found in power transmissions where heavy loads necessitate close-fitting threads. Another common application for Acme threads is in the lead screws and feed screws of precision machine tools. The cutting tool for cutting external Acme threads is shown in Fig. 9-14, and the cutting tool for internal Acme threads is shown in Fig. 9-15.

The square thread form is used for many vise and clamp screws. In cutting a square thread with a large lead, the tool angles must be absolutely correct. Clearance should be allowed on two sides, tapering from both the top and front of the tool. The tool must be fed directly into the work with the cross feed (or compound-rest

202

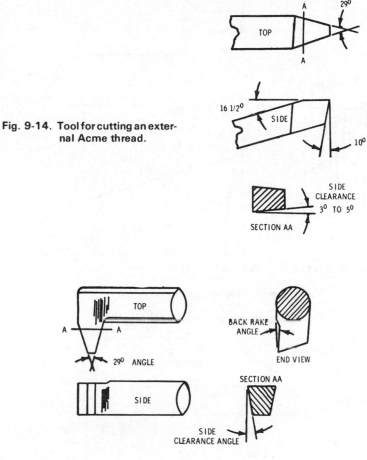

Fig. 9-14. Tool for cutting an external Acme thread.

Fig. 9-15. Tool for cutting an internal Acme thread.

feed), and care must be taken to avoid chatter and "hogging in." The tool for cutting external square threads is shown in Fig. 9-16, and the tool for cutting internal square threads is shown in Fig. 9-17.

The Whitworth thread form has been used mostly in England. Great care is required in grinding this cutting tool, so that it will produce the correct radius at both the top and bottom of the screw thread.

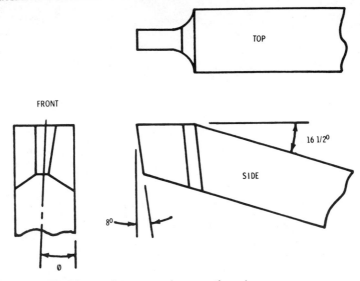

Fig. 9-16. Tool for cutting external square threads.

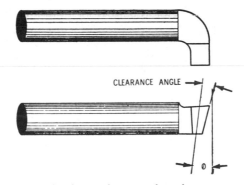

Fig. 9-17. Tool for cutting internal square threads.

THREAD CUTTING OPERATIONS

The sharp-v thread is used as an example for thread cutting. If a change-gear lathe is used, the change gears should be properly in mesh, and the operator should note that:

1. For right-hand threads, the carriage must travel toward the headstock.

2. For left-hand threads, the carriage must travel toward the tailstock.
3. Reversing the direction of rotation of the lead screw (by shifting the tumbler gears) will cause the carriage to move in the opposite direction, without changing the direction of spindle rotation.

Preliminary Setup for Thread Cutting

If a lathe with a quick change gear box is used, it is necessary only to place the tumbler levers in the proper position to cut the desired number of threads per inch, as indicated on the index plate. In addition to setting up the lathe to cut the desired number of threads per inch, the cutting tool and the work must be mounted properly.

Setting the cutting tool—For cutting external threads, the top of the threading tool should be placed exactly on center (Fig. 9-18). The tool should be set square with the work (Fig. 9-19). The center gage can be used to adjust the point of the cutting tool so that the angle of the thread will be correct.

Fig. 9-18. Set the top of the cutting tool exactly on center when cutting screw threads.

For cutting internal threads, the cutting tool should also be set exactly on center and square with the work (Fig. 9-20). The boring bar should be as large in diameter and as short in length as possible to prevent springing. Allow sufficient clearance between the threading tool and the inside diameter of the hole to permit backing out the tool when the end of the cut has been reached. More front clearance is required in cutting internal threads to prevent the heel of the cutting tool from rubbing, than when cutting external threads.

Setting the compound rest—For cutting 60° screw threads, the compound rest should be set at an angle of 29° (see Fig. 9-19). The depth of cut can be adjusted on the compound-rest micrometer dial. Most of the metal is removed by the left-hand side of the

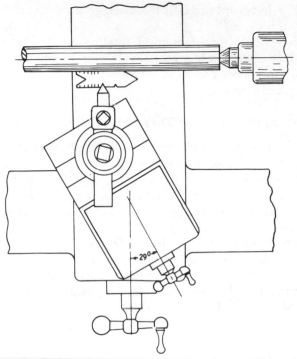

Fig. 9-19. Setup for cutting a 60° American National standard thread.

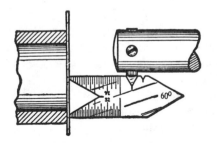

Fig. 9-20. Set the cutting tool square with the work when cutting internal screw threads.

threading tool when the compound rest is set at an angle of 29° (Fig 9-21). This permits the chip to curl out of the way and prevents tearing the thread. As the angle on the side of the threading tool is 30°, the right-hand side of the threading tool will shave the thread smoothly, leaving a fine finish, although it does not remove

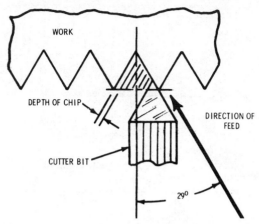

Fig. 9-21. Action of the thread cutting tool when the compound rest is set at a 29° angle.

enough metal to interfere with the main chip which is removed by the left-hand side of the threading tool.

Cutting oil for thread cutting—Either lard oil or machine oil should be used if a smooth thread is desired on steel. The oil should be applied generously before each cut to prevent tearing the steel by the cutting tool, thus causing a rough finish on the sides of the threads.

Cutting the Threads

After the lathe has been set to cut the desired number of threads per inch, and both the thread cutting tool and work have been mounted properly, a very light trial cut should be taken (Fig. 9-22). The chief purpose of the trial cut is to make certain that the desired thread pitch is being obtained.

The number of threads per inch can be checked by placing a steel rule against the work so that the end of the rule rests on the point of a thread or one of the scribed lines (Fig. 9-23). The number of spaces between the end of the rule and the first inch mark indicates the number of threads per inch (11½ threads per inch in Fig. 9-23).

Use of thread stop—The thread stop can be used to relocate the cutting tool for each successive cut (Fig. 9-24). The thread cutting tool must be withdrawn quickly after each cut to prevent the point

207

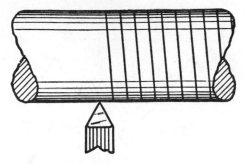

Fig. 9-22. A trial cut can be used to check the lathe setup for the correct thread.

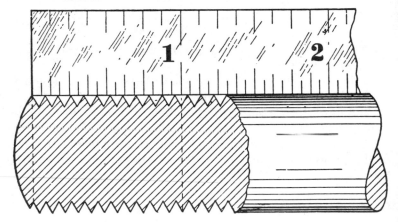

Fig. 9-23. A method of checking screw thread pitch.

of the tool from digging into the metal. It is difficult to stop the threading tool abruptly, so provision for clearance is usually made at the end of the cut. A neck or groove provides clearance for the tool to run out at the end of the cut.

The point of the cutting tool should be set so that it touches the work lightly. Then the thread stop should be locked to the saddle dovetail at about ¼ (0.25) inch from the base of the compound rest; the thread stop adjusting screw should be turned until the shoulder is tight against the stop.

To withdraw the cutting tool after each successive cut, turn the cross-feed handle several turns to the left, and return the carriage to the point where the thread is to begin. Then turn the cross-feed

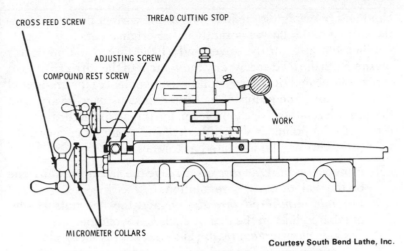

CROSS FEED SCREW

THREAD CUTTING STOP

ADJUSTING SCREW

COMPOUND REST SCREW

WORK

MICROMETER COLLARS

Courtesy South Bend Lathe, Inc.

Fig. 9-24. A thread stop attached to the dovetail of the saddle.

handle to the right until the thread cutting stop screw contacts the thread stop. This places the cutting tool in its original position. Turn inward 0.002 inch or 0.003 inch on the compound-rest micrometer collar for each successive cut.

Use of the thread dial indicator—The thread dial indicates the relative positions of the lead screw, the carriage, and the spindle of the lathe (Fig. 9-25). This permits disengaging the half-nuts at the end of a cut, returning the carriage quickly to the starting point by

Fig. 9-25. Thread dial indicator attached to the lathe carriage.

Courtesy South Bend Lathe, Inc.

hand and re-engaging the half-nuts at the correct point to assure the cutting tool following exactly in the original cut.

When the gear on the lower end of the thread dial indicator meshes with the lead screw, any movement of the carriage or lead screw is indicated by a corresponding movement of the graduated dial at the top. The points of half-nut engagement will vary with the individual lathe. For example, on the thread dial indicator (see Fig. 9-25) the points at which the half-nuts may be engaged for successive cuts will be indicated as follows:

1. *For even-numbered threads*, engage the half-nuts at any line on the dial, or each ⅛ revolution.
2. *For odd-numbered threads*, engage the half-nuts at any numbered line on the dial or each ¼ revolution.
3. *For half-numbered threads* (11 ½ threads per inch), close the half-nuts at any odd-numbered line, or each ½ revolution.
4. *For quarter-numbered threads* (4 ¾ threads per inch), return to the original starting point for each successive cut.

Resetting the threading tool—If it is necessary to remove the threading tool before the thread has been completed, the tool must be carefully adjusted to follow the original groove when it is replaced in the lathe. All lost motion can be taken up by rotating the drive belt by hand. By adjusting the compound-rest and the cross-feed screws simultaneously, the threading tool can be adjusted to enter the original groove.

Fitting and checking threads—A screw thread pitch gage can be used to check the finer pitches of threads (Fig. 9-26). Either the nut that is to be used on the thread or a ring type of thread gage can be used to check the thread (Fig. 9-27). The nut should fit snugly

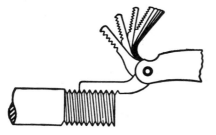

Fig. 9-26. A screw thread pitch gage.

Courtesy South Bend Lathe, Inc.

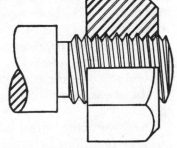

Fig. 9-27. Fitting a screw thread to a nut to check fit.

Courtesy South Bend Lathe, Inc.

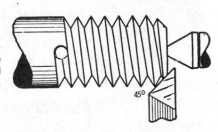

Fig. 9-28. A 45° chamfer is commonly used to finish the end of the screw thread on bolts, cap screws, and so forth.

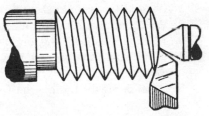

Fig. 9-29. Finishing the end of a thread by rounding with a forming tool.

without any play, but it should not bind on the thread at any point.

Finishing the end of a thread—A 45° chamfer is commonly used to finish the end of a thread on bolts, cap screws, etc. (Fig. 9-28). A forming tool is often used to round the ends of machines parts and special screws (Fig. 9-29).

SUMMARY

The most common form of a screw thread is one that has symmetrical sides inclined at equal angles to a vertical center line through the apex of the thread. The American National screw thread was the standard thread used in the United States for many years but now has been replaced by the Unified, Whitworth, and Acme forms.

Diameter pitch combinations are grouped to form thread series which are distinguished by the number of threads per inch as applied to a specific diameter. There are eleven standard series combinations, such as coarse-thread, fine-thread, and extra-fine thread.

The shape of the cutting tool depends on the type of thread to be cut. The tool must be ground and set accurately or the correct thread form will not be obtained. Various gages are available that can be used as guides in grinding. The Whitworth thread form has been used mostly in England. Extreme care is required in grinding this cutting tool so that it will produce the correct radius at both the top and bottom of the screw thread.

REVIEW QUESTIONS

1. Name the various screw thread forms.
2. What country uses the Whitworth thread form?
3. What is meant by the angle of thread?
4. What standard thread was used for years in the United States?
5. What is meant by length of engagement?

CHAPTER 10

Lathe Attachments

The many attachments that are available for the engine lathe make it a very versatile machine. Some of these attachments facilitate the usual lathe operations, and others permit a variety of machining operations that are not strictly lathe work.

WORK SUPPORT ATTACHMENTS

In many lathe operations the work cannot be held between the lathe centers, and an attachment must be used to aid in supporting the work. This is especially true when it is necessary to bore, drill, thread, or perform other similar internal operations on the end of a long piece of work.

Follower Rest

The follower rest is used to support long slender stock to prevent its springing away from the cutting tool. The follower rest is bolted to the saddle of the carriage; therefore, it "follows" the cutting tool. The frame of the follower rest is shaped like a question mark, and it has both vertical and horizontal adjustable jaws (Fig. 10-1).

The jaws of the follower rest form a true bearing for the work, permitting it to turn without binding. In setting the follower rest jaws, first remove the guard over the cross feed screw—place a small piece of paper over the screw to keep off chips during the cutting operation; the dovetail ways should be wiped clean. Adjust the jaws with the carriage near the tailstock after a short portion of

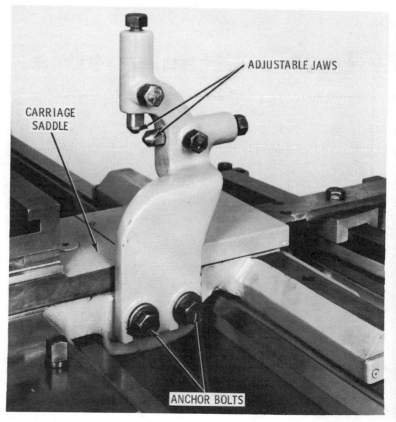

Courtesy Cincinnati Milacron Co.

Fig. 10-1. A follower rest for a lathe.

the work has been turned. Set the vertical jaw to touch the top of the workpiece. A small piece of cellophane can be placed between the work and the jaws to determine the proper amount of friction. During the cutting operation, plenty of lubricant should be applied to the workpiece where it contacts the jaws of the follower rest. The jaws should be adjusted after each cut in order to retain accuracy.

Steady Rest

The steady rest is used to support one end of the long, slender stock when the other end is centered in a chuck and the operation being performed is such that the tailstock cannot be used (Fig. 10-2). The frame of the steady rest is circular in shape and is fastened to the bed ways of the lathe (Fig. 10-3). The adjustable jaws form a bearing for the work, holding it in centered position.

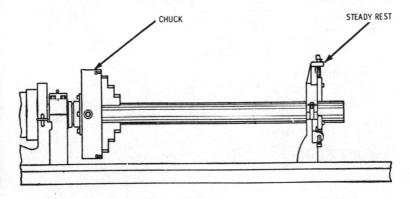

Fig. 10-2. Work mounted with one end centered in a chuck and the other end supported by a steady rest.

Positioning the jaws properly is essential when mounting the workpiece. The jaws must form a true bearing for the workpiece, permitting it to turn freely without binding or play. With one end of the work mounted in the chuck, slide the steady rest close to the chuck jaws, tighten the base clamp, adjust the jaws of the steady rest, and lock them in position on the work. A small piece of cellophane placed between the jaws and the work can be an aid in

215

Fig. 10-3. A steady rest for a lathe.

obtaining the proper bearing. Advance each jaw of the steady rest until it barely touches the work, and remove the cellophane. After tightening both the lock nut and the clamp screw on each jaw, loosen the base clamp; slide the steady rest into proper position; and retighten the base clamps. If extreme accuracy is required, trueness of the work should be checked with a dial gage.

In machining the end of a long cylindrical piece, it is preferable to secure one end of the workpiece in a chuck (see Fig. 10-2).

However, heavy belt lacing can be bound around a lathe dog to hold it to a faceplate so that the end of the work can be mounted on a headstock center (Fig. 10-4). This is a kind of makeshift setup that can be used by a person with insufficient skill for mounting the work in an independent chuck with precision.

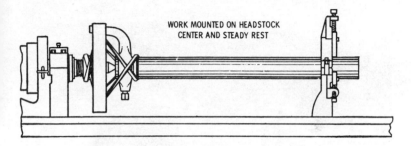

WORK MOUNTED ON HEADSTOCK
CENTER AND STEADY REST

Fig. 10-4. Belt lacing can be used to mount the work on a headstock center and steady rest.

AUTOMATIC FEED STOPS

Automatic devices that disconnect the power feed after a predetermined distance of travel are often used on lathes in production work. These devices can be used to enable a single machinist to operate two or more lathes simultaneously. The operator can place the work in a lathe, engage the power feed for a cut, and proceed to the next lathe. When the end of the cut is reached, the power feed is automatically disconnected.

Carriage Stops

The *micrometer carriage stop* is especially desirable for production work where long longitudinal cuts are required. It is indispensable for accurate control of length of cut. The fixed stop can be located at any position along the bed ways; it can be fixed at any position for repetitive work. The micrometer collar is graduated in thousandths of an inch (Fig. 10-5). The micrometer carriage stop can also be used to face shoulders to an exact length.

The *four-position length and depth stop* is designed for production turning and for facing multiple diameters and shoulders (Fig.

217

10-6). This turret-type stop has four individually adjusted stop screws. The length stop mounts on the front bed ways, and the depth stop mounts on the carriage wings and bridge. A handy *length measuring attachment* can be used to measure either the forward or the reverse movement of the lathe carriage within 0.001 inch (Fig. 10-7).

Courtesy Cincinnati Milacron Co.

Fig. 10-5. A micrometer carriage stop attachment for a lathe.

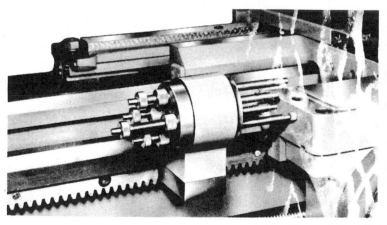

Courtesy Cincinnati Milacron Co.

Fig. 10-6. A four-position length and depth stop attachment.

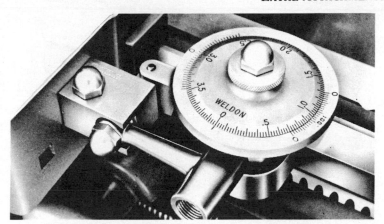

Fig. 10-7. A length-measuring attachment used to measure the forward or reverse carriage movement to 0.001 inch.

Cross-Slide Stop

The ball-type cross-feed stop enables the operator to retract the cross slide at the end of a cut and to return quickly to its original setting prior to the next cut (Fig. 10-8). When the thumbscrew is loose, the cross-feed screw can be used normally for full length of travel of the cross slide. The stop can be set quickly by adjustment of the thumbscrew.

OTHER ATTACHMENTS

A number of machining attachments are available to increase the capability of the lathe for performing a variety of operations.

Taper Attachment

A typical taper attachment for a lathe is shown in Fig. 10-9. This is probably the most widely used method of machining tapers on the engine lathe. It is much preferred to the tailstock setover method for machining large tapers. Tapers up to 4 inches per foot (20°) can be machined with the taper attachment.

Most taper attachments consist of a bar at the back of the lathe.

219

Fig. 10-8. Cross-feed stop.

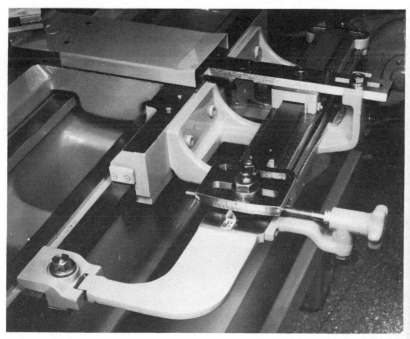

Fig. 10-9. A lathe taper attachment.

The bar is engaged by the cross slide, which moves the cutting tool transversely closer to or farther from the lathe center axis as the cutting tool travels longitudinally along the lathe bed. Duplicate tapers can be cut quickly on pieces of different lengths; and taper boring, which is impossible with the tailstock set over, can be performed easily with the taper attachment.

The swivel taper bar (L in Fig. 10-10) is graduated in both degrees J and inches per foot N, and represents the included angle or taper. The taper bar moves only one-half the amount indicated on the scale.

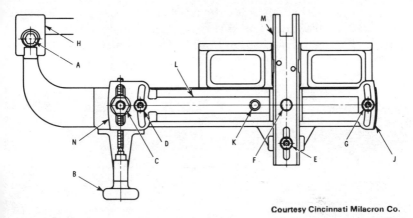

Courtesy Cincinnati Milacron Co.

Fig. 10-10. Diagram of a typical taper attachment showing parts.

In adjusting the taper attachment (see Fig. 10-10), the cross-slide bar M should be kept centrally located on the swivel taper bar for easy adjustment (keep the screw F in the cross-slide bar over the stud K in the taper bar). To set the taper attachment for a desired taper, loosen the clamping nuts C, D, E, F, and G. Then the swivel-taper bar can be adjusted to the approximate setting by hand. A final adjustment can be made by tightening the nut C, and using the adjusting screw B. Position the cutting tool so that the taper attachment will travel the full length of the part to be turned. Then tighten the engaging nut A to engage the taper attachment.

The attachment can be disengaged by loosening the engaging nut A and tightening the clamping nut E, which must remain tight except during the tapering operation. The clamping block H should be removed if the taper attachment is not used frequently, to reduce wear on the bed ways.

The taper attachment can also be used to bore a tapered hole. Set the attachment in the same manner as for turning external tapers, except that the angle of the taper bar is usually reversed. Make certain that the tool will clear the smallest diameter of the tapered hole and that the tool is set on center to avoid boring either a concave or a convex surface onto the workpiece.

Tracer Attachment

The tracer attachment for the lathe can be used for quick and accurate duplication of parts (Fig. 10-11). It can be set to cut taper angles between 0° and 90° in both facing and turning operations. The template is positioned on the longitudinal template support bar.

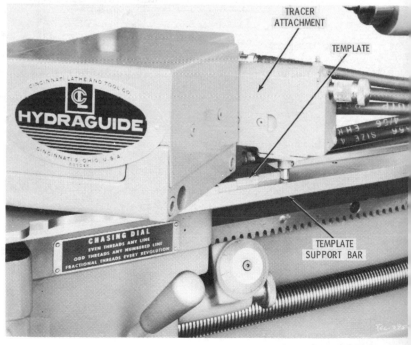

Fig. 10-11. Tracer attachment for a lathe.

The hydraulic power slide has a normal movement angle of 45° to the center line of the lathe. Maximum power slide travel is 3¾ (3.750) inches. This corresponds to a movement of 2⅝ (2.625) inches perpendicular to or parallel to the center line of the work.

Milling Attachment

This attachment can be used for milling work in a small shop that cannot afford to install a milling machine (Fig. 10-12). To install the milling attachment on the lathe, remove the compound rest, and clamp the base of the milling attachment in its place. The milling attachment equips the engine lathe for face milling, cutting keyways and slots, milling dovetails, squaring shafts, and making dies, molds, etc.

The vise position is controlled by a feed screw with a graduated micrometer collar. The attachment can be swiveled to hold work at any angle.

Fig. 10-12. Milling attachment for a lathe.

223

The cutting speed for a milling operation should be approximately two-thirds the speed recommended for turning the material in the lathe. The cut is controlled by the lathe carriage handwheel, the lathe cross-feed screw, and the vertical adjusting screw at the top of the milling attachment. All milling cuts should be taken with the direction of cutter rotation against the direction of feed of the workpiece.

Gear Cutting Attachment

Spur and bevel gears of all kind can be cut on the lathe gear cutting attachment. The attachment equips the lathe for graduating and milling, external key-seat cuts, angle cuts, spline cuts, slotting work, and regular dividing head milling work.

The device is designed for mounting on the milling vise. The cutter is mounted on a mandrel which turns on the lathe centers.

Grinding Attachment

A tool-post grinder can be mounted in the tool slide of the lathe compound rest (Fig. 10-13). This attachment equips the lathe for

Fig. 10-13. Tool-pot grinding attachment for a lathe.

both internal and external precision grinding (Fig. 10-14). It can be used to grind reamers and milling cutters, dies, gages, bushings, bearings, shafts, valves, and valve seats. The grinding attachments are driven by small electric motors—the lathe spindle must be reversed for grinding operations (see Fig. 10-13); and as the grinders are constructed for attachment to the compound rest, the grinding wheel can be adjusted to the proper position for the various grinding operations. As grinding is a finishing operation, the work should be turned as near the finished sizes as possible before beginning the operation.

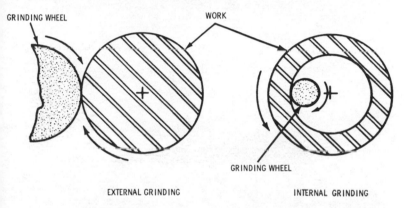

EXTERNAL GRINDING   INTERNAL GRINDING

Fig. 10-14. Diagram illustrating external (left) and internal (right) grinding on the lathe. The work must always turn in a direction opposite that of the grinding wheel. For external grinding, it is necessary to reverse the lathe spindle. For internal grinding, the lathe spindle should be run in the forward direction.

Turret Attachments

Turrets can be used on the lathe bed, tailstock, and tool post. Advantages of these attachments are elimination of second-operation setups and rapid and accurate machining of duplicate parts on a production basis.

Bed turret—The bed turret is a cylinder or head arranged to turn and slide on the lathe bed ways (Fig. 10-15). It can be fitted with various tools for boring, reaming, tapping, etc. The head indexes

225

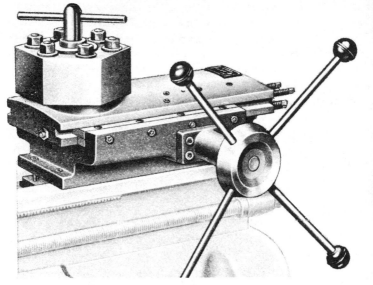

Fig. 10-15. A lathe bed turret.

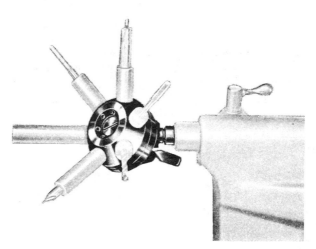

Fig. 10-16. A lathe tailstock turret.

⅙ of a turn with each complete backward movement of the handwheel.

Tailstock turret—This device mounts in the lathe tailstock spindle (Fig. 10-16). Any of the six tool stations can be selected quickly by a convenient trip lever control. The attachment can be used to speed short-run work and to eliminate the need for second-operation setups.

Tool-post turret—This attachment can be mounted on the compound rest (Fig. 10-17). Various cutting tools can be mounted in the turret for facing, turning, etc. A turn of the handle releases the head for rotating to the next operating position.

Fig. 10-17. A tool-post turret.

Courtesy Atlas Press Co.

SUMMARY

Many lathe attachments are available that make the lathe a very versatile machine. In many lathe operations the work cannot be held between the lathe centers, and an attachment must be used to aid in supporting the steady rest. The follower rest prevents long slender stock from springing away from the cutting tool. The steady rest supports stock when the other end is centered in a lathe chuck and the tailstock cannot be used.

Automatic devices that disconnect the power feed after a predetermined distance of travel are often used on lathes in production work. With this device, a single machinist can generally operate two lathes simultaneously. He can place the work on one lathe, engage the power feed for a cut, and proceed to the next lathe. When the end of the cut is reached, the power feed is disconnected automatically.

REVIEW QUESTIONS

1. What is a follower rest?
2. What is a steady rest?
3. What other attachments are available for various lathe operations?
4. What is an automatic feed stop?
5. Where is the micrometer carriage stop used in production work?
6. What is the purpose of the ball-type cross-feed stop?
7. Where is belt lacing used, and why?
8. What is taper?
9. How are tapers cut on a lathe?
10. What is a tracer attachment?

CHAPTER 11

The Turret Lathe

The turret lathe is a comparatively modern machine that has been developed from the engine lathe by the addition of revolving tool-holding devices called turrets. The turret or head can be arranged to turn on an upright axis and to slide on the ways of the lathe; it is fitted with holes that hold cutting tools, any of which can be applied as needed in the axial line of the work by turning the head.

At first, turrets were made in circular form and were rotated on a vertical pivot that could be held in any desired position by a binding nut. The periphery of the circular turret was drilled and reamed to hold four tools, which projected from the turret at 90° angles with each other. Later, the turret was made in hexagonal form, and the number of tools held in the turret was increased to six.

The earlier turrets were mounted on the lathe carriage in place of the tool block, and were aligned with the lathe axis by means of the cross-feed screw. Lateral feed was obtained from the feed mechanism in the apron attached to the carriage.

The chief objective accomplished by addition of the turret to the earlier lathes was that it enabled the operator to perform drilling, reaming, counterboring, and similar operations on a piece of work without a change of tools other than revolving the turret. After they had been set and adjusted, the tools in the turret required no further alterations as the workpieces were completed and removed from the chuck and other pieces substituted for a like series of operations. Thus, work could be performed more rapidly by this means. The turret mechanism was developed further by the addition to the number of tools it could carry, by a ratchet arrangement for revolving the turret, by an index plate for holding the turret in any desired position, and by various other improvements in design.

All these improvements led to the design of special lathes in which the improved turret was a special feature, and led to the development of the turret lathe as it is built today.

The turret lathe is one of the most useful machines in the machine shop for the production, in large quantities, of work that requires repeated operations, such as turning, boring, or reaming. The turret is equipped with the appropriate tools, and the work, such as a casting or forging, is held by a chuck.

CLASSIFICATION

Turret lathes are classified as either horizontal or vertical. In general, unless specified, a turret lathe is understood to be a horizontal turret lathe. Horizontal turret lathes are classified as either bar machines or chucking machines.

Bar machines are used to machine bar stock or castings that are about the same size and shape as bar stock. The chucking machines are used to machine castings and forgings that must be held in chucks or fixtures. Bar and chucking machines are classified further as either ram-type turret lathes or saddle-type turret lathes.

230

Ram Type

On the ram-type turret lathe, the turret is mounted on a slide, or ram, which moves back and forth on a saddle clamped to the lathe bed (Fig. 11-1). This machine is a fast and easily operated machine for small work. It is quickly and easily operated because the hexagonal turret can be moved back and forth without moving the entire saddle unit. The machine has a short turret stroke and an automatic index on the turret. The ram-type turret lathe is well suited for bar work and chucking work where the overhang of the ram can be kept short.

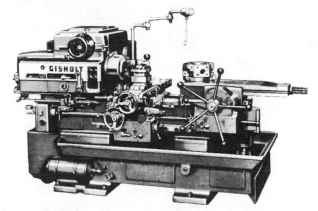

Courtesy Gisholt Machine Co.

Fig. 11-1. Universal ram-type turret lathe.

Courtesy Gisholt Machine Co.

Fig. 11-2. Saddle-type turret lathe.

231

Saddle Type

On the saddle-type turret lathe, the turret is also mounted on the saddle, but the entire saddle unit moves back and forth on the lathe bed (Fig. 11-2). This machine is more heavily constructed than the ram type and is more suitable for work requiring long turning and boring cuts. It has a long turret stroke and a rigid turret mounting.

FUNDAMENTAL PARTS

The basic parts of a turret lathe are: headstock, ram saddle, turret, cross-slide carriage, and chuck. In addition, various automatic devices can be used on turret lathes.

Headstock

One of the most important operating units of any turret lathe is the headstock. Either an electric head with a multiple-speed motor mounted directly on the spindle or an all-geared head is used on the modern turret lathe. These heads provide a wider range of speeds and permit heavier cuts than were possible on the earlier cone-drive heads. The operator needs only to set the dial to the diameter of the work (or the desired speed), and the spindle speed selector will automatically shift to the correct speed.

Turret Ram and Saddle

In the ram-type turret lathes, the ram provides a base for the turret (Fig. 11-3). The turret is automatically unclamped and indexed by backward movement of the ram. Forward movement clamps the turret before it leaves the saddle. The turret is free for backing up or skip indexing at one point in the travel of the ram. The longitudinal feed handwheel moves the ram in the saddle and permits leverage when feeding by hand from a normal working position.

There are three general classes of rams based on the movement or movements of the turret. The three classes of rams are:

1. Plain.
2. Offset.
3. Universal.

The plain slide, or ram, provides only longitudinal movement; the turret is rotated automatically to bring the next tool into position on the backward or return stroke.

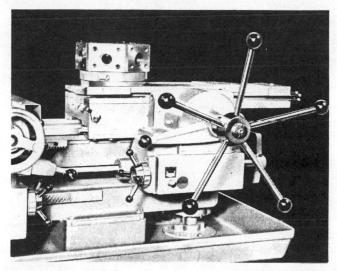

Courtesy Gisholt Machine Co.

Fig. 11-3. Turret ram and saddle for a turret lathe.

The setover, of offset slide, provides both longitudinal and transverse movements. The transverse motion can be used for radial facing or recessing with a single-point tool.

The universal slide, or ram, has both the longitudinal and transverse movements; and, in addition, it has an arrangement consisting of an intermediate plate between the cross slide and ram to permit swiveling for such work as turning or boring of tapered surfaces.

The three aforementioned types of rams are diagrammed in Fig. 11-4. The saddle-type turret lathe is a form of universal machine in which the turret is mounted on a swivel placed on the saddle (Fig. 11-5).

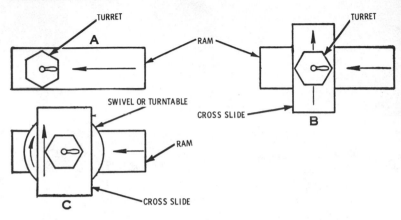

Fig. 11-4. Diagram showing three general classes of rams found on turret lathes: (A) plain, (B) offset, and (C) universal.

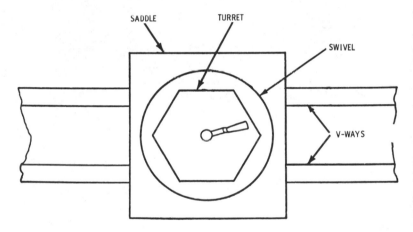

Fig. 11-5. Diagram showing the mounting of the turret on a saddle-type turret lathe.

Turret

The hexagon-shaped turret in the ram-type turret lathe is mounted on a slide, or ram, which moves back and forth in the saddle (see Fig. 11-3). The saddle is clamped to the bed ways in proper position for the job to be performed.

234

Square turrets on ram-type turret lathes are adjustable along the slide to obtain proper tool clearances. The square turret is mounted directly on the slide in saddle-type turret lathes.

Cross-Slide Carriage

The cross-slide carriage is held to the bed ways by flat hold-down plates with tapered gibs for wear adjustment (Fig. 11-6). Eight selective, reversible power cross and longitudinal feeds are available on the cross-side carriage shown in Fig. 11-6.

A dial-type feed selector gives a direct reading in thousandths per revolution for each of the eight power feeds. A lever at the right of the apron engages the cross feed, and a lever at the lower left front of the apron selects either forward or reverse feed.

Courtesy Gisholt Machine Co.

Fig. 11-6. Cross-slide carriage on a ram-type turret lathe.

Chuck

A collet chuck (Fig. 11-7) is commonly found on turret lathes. It is hydraulically operated and controlled by movement of the "chuck-unchuck" lever. Round, square, or hexagon bar stock can be handled easily with this type of chuck.

Fig. 11-7. A hydraulically operated collet chuck. It is controlled by movement of the "chuck-unchuck" lever.

Courtesy Gisholt Machine Co.

Courtesy Gisholt Machine Co.

Fig. 11-8. Four-position stop-roll attachment.

Turret Lathe Attachments

Numerous automatic devices are used on the modern turret lathe to control turret movements that bring the tool into position at the correct moment and to feed the work through the chuck after each piece is finished. Various "stops" are placed on top of the lathe bed—not to disengage the feed, but to arrest the travel of the carriage while the feed continues to hold the carriage firmly against the stop.

A *four-position stop roll* (Fig. 11-8) can be used with both front and rear cross-slide rolls. The stop roll is hand indexed, and it can be set up quickly to trip cross feeds in either direction. Each feed trip dog is equipped with an independently adjustable dead stop to permit hand feeding to an exact size, while referring to a dial or indicator, after the power feed has been disengaged. Also, the dead stop can be used for accurate setting of the cross slide for turning operations.

A *taper attachment* for the universal carriage is a separate tool bolted to the T-slots on the rear of the cross slide (Fig. 11-9). The rear tool block can be used for taper turning without affecting normal use of the square turret at the front of the cross slide.

Courtesy Gisholt Machine Co.

Fig. 11-9. Taper attachment for the universal carriage.

A *rapid traverse cross-slide attachment* (Fig. 11-10) reduces operator effort and speeds machining operations. An independent electric motor drive unit is mounted on a bracket at the rear of the cross slide and connected by gears to the cross-feed screw. Manual operation of a switch control provides traverse in either direction.

Fig. 11-10. An electric rapid traverse attachment to the cross slide of a turret lathe.

SUPER-PRECISION TURRET LATHES

The bench turret lathe operates in much the same manner as the larger turret lathes. This type of machine tool is designed for short-run and long-run super-precision secondary machining operations that require extremely close tolerances.

Figs. 11-11 through 11-39 illustrate the bench turret lathe as well as a wide variety of features, attachments, and operations performed on this precision machine tool.

SUMMARY

Turret lathes are classified as either horizontal or vertical. In general, a turret lathe is understood to be a horizontal type. Turret lathes are usually classified as either bar machines or chucking machines. The chucking machine is used to machine castings and forgings that must be held in chucks or fixtures. The bar machine is used to machine bar stock or castings that are about the same size

Fig. 11-11. This bench turret lathe is a super-precision second-operation machine for machining extremely close tolerances.

Fig. 11-12. A drilling operation.

Fig. 11-13. An internal turning operation.

Fig. 11-14. An operation using a cutoff tool.

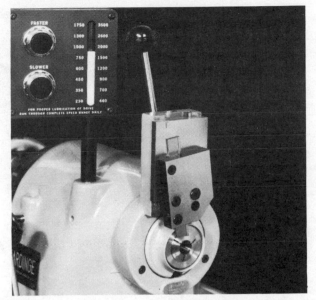

Fig. 11-15. Vertical cutoff slide.

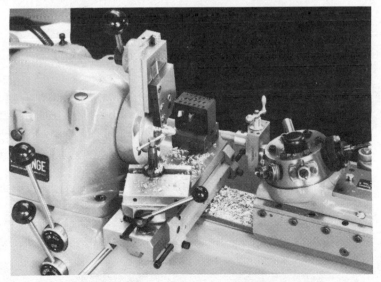

Fig. 11-16. A cutoff operation using a vertical cutoff slide.

Fig. 11-17. A boring operation involving a tapered hole.

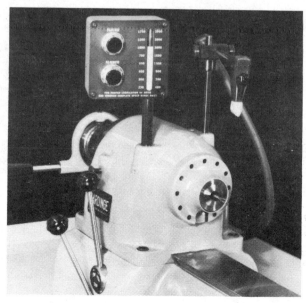

Fig. 11-18. A collet for holding parts.

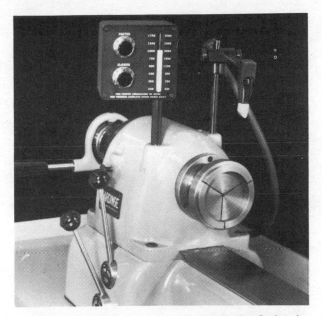

Fig. 11-19. A step chuck for holding parts.

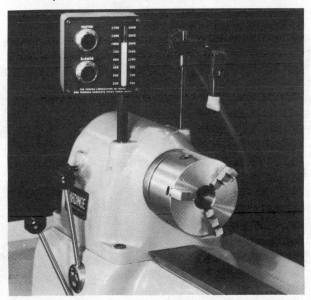

Fig. 11-20. A three-jaw chuck for holding parts.

Fig. 11-21. An internal chucking operation.

Fig. 11-22. ''Slower'' button is pressed to decrease the spindle speed.

Fig. 11-23. "Faster" button is pressed to increase the spindle speed.

Fig. 11-24. The double-tool cross slide.

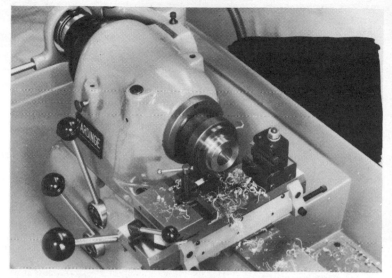

Fig. 11-25. Preloaded ball-bearing turret. Specially designed for speed and ease of operation.

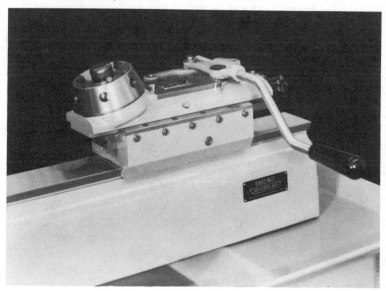

Fig. 11-26. A turning operation using a step chuck.

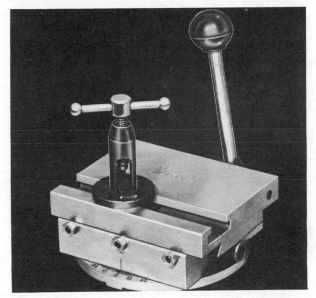

Fig. 11-27. Straight and taper turning slide for double-tool cross slide.

Fig. 11-28. High-production precision turning of stainless steel carburetor needle valves.

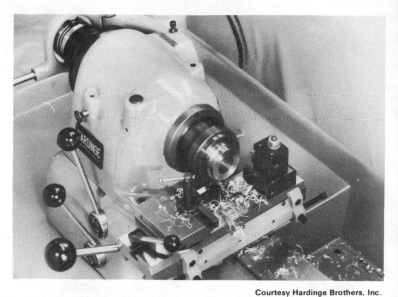

Courtesy Hardinge Brothers, Inc.

Fig. 11-29. Slide set for straight turning of a large-diameter aluminum part.

Courtesy Hardinge Brothers, Inc.

Fig. 11-30. Slide setup in front position of the double-tool cross slide for precision turning of a taper on a brass part.

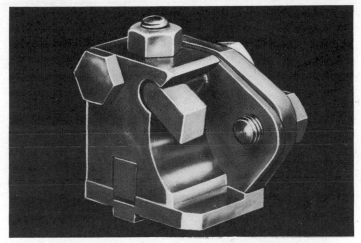

Fig. 11-31. Circular form toolholder.

Fig. 11-32. The holder fits directly to the double-tool cross-slide toolholder
block.

Fig. 11-33. Multiple toolholder.

Fig. 11-34. With the multiple toolholder, many operations such as under-
cutting, chamfering, and grooving can be done in one operation.

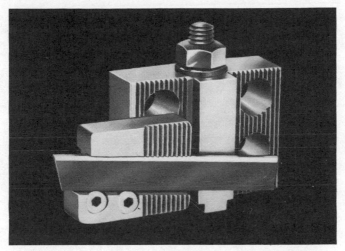

Fig. 11-35. Cutoff toolholder.

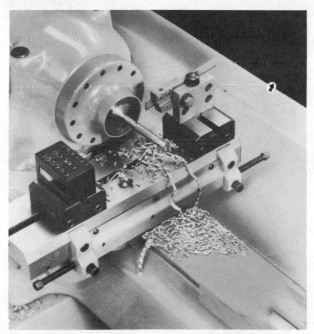

Fig. 11-36. The cutoff toolholder fits directly to the front or rear tool block
on the double-tool cross slide.

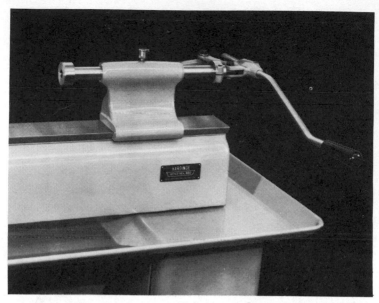

Fig. 11-37. This slide is for deep hole drilling, long threading, box tool turning, lapping, or any other operation requiring tool travel up to 5½ inches.

Fig. 11-38. Deep hole drilling.

Courtesy Hardinge Brothers, Inc.

Fig. 11-39. Production threading.

and shape as bar stock. Bar and chucking machines are further classified as ram-type turret or saddle-type turret lathes.

On the ram-type lathe, the turret is mounted on a slide, or ram, which moves back and forth on a saddle that is clamped to the lathe bed. On the saddle lathe, the turret is also mounted on the saddle, but the entire saddle unit moves back and forth on the lathe bed.

The basic parts of a turret lathe are headstock, ram saddle, turret, slide carriage, and chuck. Various automatic devices can be used on the turret lathe, such as stops placed on top of the lathe bed to limit the travel of the carriage while the feed continues to hold the carriage firmly against the stop.

REVIEW QUESTIONS

1. How are turret lathes classified?
2. What is a chucking machine?
3. What is a bar machine?
4. What are the advantages of a turret lathe over an engine lathe?

CHAPTER 12

Automatic Lathes

Automatic lathes are ideally suited for medium to long production runs. In many instance, the automatic cycle enables the operator either to handle additional units or to perform other work.

AUTOMATIC TURRET LATHES

The automatic turret lathe is especially suited to small-lot and long production runs (Fig. 12-1). All functions of this lathe are under complete computer numerical control so that the primary functions of the operator consist of loading the machine and inspecting the finished part. The lathe is a universal turning center designed for bar, chucking, and shaft work.

Courtesy Sheldon Machine Co.

Fig. 12-1. This computer numerical-controlled lathe is a Universal turning center designed for bar, chucking, and shaft work.

Illustrated are many different parts in a variety of shapes, sizes, and types of material (Fig. 12-2). The lathes have one thing in common. They can perform several operations to produce uniform and accurate parts in large quantities with little scrap.

In order to get from "print to finished" part, a number of steps have to take place (Fig. 12-3 and 12-4). The control system enables

Courtesy Sheldon Machine Co.

Fig. 12-2. Many different shapes and sizes of parts produced from different types of materials.

the operator to increase his efficiency. The control system also has a computer retrieval terminal to display operating programs, tool compensation, commands, positions, and program edits. (Fig. 12-5). The automatic turret lathes also have a number of options to increase the capability of the lathe (Fig. 12-6 and 12-7).

257

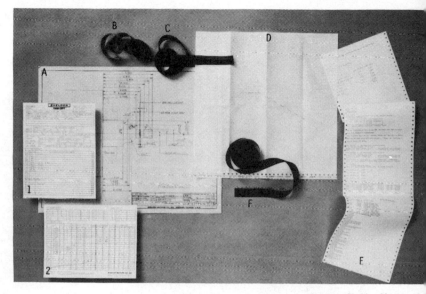

Courtesy Sheldon Machine Co

Fig. 12-3. From the dimensioned print (A) the programmer obtains necessary information to fill in forms 1 and 2. The first form allows for
general comments and tooling, while the second contains the
numerical data describing the rough stock and the finished part.
These forms are then given to a typist who, with the use of master
tape (B), fills in the information written by the programmer to
produce input tape (C). This tape is transmitted to Sheldon by
dataphone and processed by computer. With this information the
computer generates the complete machining cycle. It provides a
plot showing the paths of each tool (D); complete printout,
including the time to produce the part (E); and the final output
tape (F). This tape contains the setup instructions and operating
program and is transmitted to the user by a dataphone terminal.

Courtesy Sheldon Machine Co.

Fig. 12-4. Tape containing the setup instructions and operation program is transmitted by dataphone terminal.

AUTOMATIC THREADING LATHES

All types of internal and external threads can be produced on the automatic threading lathe (Fig. 12-8). These machines can produce threads in much less time than threads can be produced either by thread grinding or by thread milling. The supporting saddle of the automatic threading lathe is moved along the bed to position the tool in relationship to the work. The saddle is clamped during threading, which is performed by a longitudinal slide and

259

Fig. 12-5. These controls give the operator complete information about the status of the lathe and the machining operation.

cross slide mounted on the saddle. The slides are cam controlled and provide a very accurate thread at a high rate of production.

SUMMARY

For long production runs, the automatic turret lathe is an ideal piece of machinery. As an example, the functions of the lathe are

Courtesy Sheldon Machine Co.

Fig. 12-6. Chuck with collet pads shows bar feed application. Simply by changing jaws the lathe can be converted to either a bar or chucking application.

under complete computer control so that the primary functions of the operator consist of loading the machine and inspecting the finished part.

The automatic turret lathe is capable of producing parts in a variety of sizes, shapes, and types of material. They can perform several operations to produce uniform and accurate parts in large quantities with little scrap. All types of internal and external threads can be produced on the automatic threading lathe in much less time than can be produced either by thread grinding or by thread milling.

261

Fig. 12-7. Optional, 8-station automatic indexing front turret permits total of 12 tools to be used in one setup. This helps eliminate tool changes from job to job. This turret uses standard qualified 1" × 1" × 4" tools.

Fig. 12-8. Automatic threading lathe.

REVIEW QUESTIONS

1. Why are automatic lathes important in long production runs?
2. What are some of the advantages of an automatic threading lathe?

CHAPTER 13

The Automatic Screw Machine

Because of modern developments, the term *screw machine* is practically a misnomer. The original screw machine was developed many years ago. Development of the original screw machine, by the addition of automatic devices, has resulted in machines being designed for various applications, each having it own field.

The screw machine was designed originally for making small screws and studs; the flexibility of the machine has adapted it to a large variety of work, resulting in specialized types of machines adapted to other operations.

A screw machine is a type of turret lathe so designed that it rapidly positions a series of tools to the piece of stock held in the chuck for a series of operations to be performed on it. The screw machine differs from the lathe in that it is provided with a front and rear cross slide, and the tailstock is replaced by a slide with a turret mounted on it. The turret is provided with several tool positions, which can be indexed into a working position as the slide is advanced, permitting each successive tool to perform an operation on the part being turned.

CLASSIFICATION

The numerous types of screw machines can be classified as follows:

1. Type of operation.
   a. Plain or hand operated.
   b. Semiautomatic.
   c. Automatic.
2. Type of spindle.
   a. Single.
   b. Multispindle.

Drilling, boring, reaming, forming, facing, etc., are examples of operations that can be performed on the screw machine. In a semiautomatic screw machine, the rough piece of work must be placed in the chuck by the operator; the various operations, including rotation of the turret to bring the tools into position, are performed automatically.

In the automatic screw machine, the rough casting or drop forging is placed in a hopper or magazine. Then they pass to the chuck, which grips them automatically for the cycle of machining operations. The automatic cycle can be compared with that of a boiler and automatic stoker in which the only attention required of the operator is to keep a supply of coal available.

A single-spindle screw machine feeds a single piece of stock that is machined by each tool in succession in the indexing turret. A multiple-spindle machine feeds a number of pieces of stock progressively, that is, the spindles index so that the stock in each spindle is fed progressively to each tool. The indexing operation is performed automatically on semiautomatic and automatic machines. The indexing turret is one in which an appropriate gear is turned through equal angles to lock and bring each tool into working position.

OPERATING PRINCIPLES

The general principles of automatic screw machines are relatively simple. The single-spindle machine is referred to here as an example (Fig. 13-1.)

In the machine shown in Fig. 13-1, the length of bar stock feed is preset by a simple hand crank adjustment and is indicated on a scale bar. A feed shell, attached to a feed tube, is used to grip the bar stock. The feed tube is mounted in a cartridge and attached to the feeding member through a latch; thus, a quick method for changing feed shell pads is provided (Fig. 13-2). The feeding member is operated by a cam mounted on the main camshaft.

Standard cams control the forward and return motions of the turret, as well as that of the two cross slides, and the independent cutoff attachment. Turret feeds can be varied by adjusting standard cams on the regulating wheel; and cross-slide feeds are a proportion of the turret feeds. The cross-slide cams can be set relative to any turret operation by adjusting the cross-slide cam drum on the main cam shaft (Fig. 13-3).

Cutting lubricant must be used frequently to avoid excessive heat and to prevent ruining either the work or the tools. The

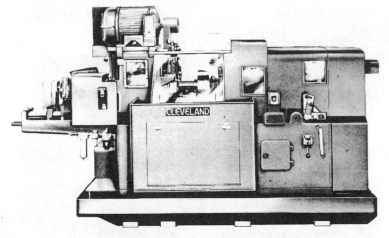

Courtesy Cleveland Automatic Machine Co.

Fig. 13-1. A single-spindle automatic screw machine.

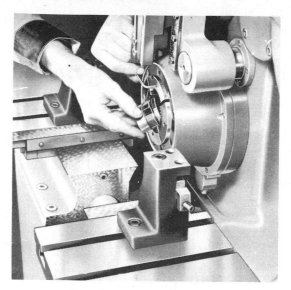

Fig. 13-2. Collet pads can be changed without removing the master collet from the chuck.

Fig. 13-3. Front and rear cross slides. Cross-slide tools can be adjusted precisely in relation to turret tools.

operating parts should be kept clean and free of any gummy coating of dried lubricant, chips, or dirt. If this is neglected, the machine can become clogged to an extent that will interfere with its proper operation. The machine must be well cared for as exacting demands are made on the machine and it is expected to run continuously and to work within extremely close limits.

SELECTION AND USE OF TOOLS

The quality of work produced by the automatic screw machine is determined by the proper selection and use of tools; these factors also are related to the accuracy, finish, and speed with which the work is turned out. The tools also must be properly ground and correctly set for best results.

When setting up a job, experience in "tooling up" is advantageous and of great help on either special jobs or very difficult jobs. However, tool principles are relatively simple; and even if the machinist is unfamiliar with a particular type of machine, he should be able to obtain the desired results by combining proper care with studying the instructions.

Types of Tools

Many different kinds of tools are used because a wide variety of work is performed on automatic screw machines. Usually, selection of the type of tool for an operation is governed by experience, although certain principles and suggestions can be follows (Fig. 13-4).

External turning tools—Balance turning tools, plain hollow mills, adjustable hollow mills, box tools, swing tools, and knee tools are all used in the turret. Tools suitable for operations on the end of the work are centering and facing tools, pointing tools, and pointing-tool holders with circular tools.

Internal cutting tools—Drills, counterbores, and reamers are held in drill or floating holders. Recessing swing tools are also used for internal operations.

Threading tools—Taps and dies, chasers mounted in opening die holders, and thread rolls held in cross-slide knurl holders or in knurling swing tools are the threading tools most generally used.

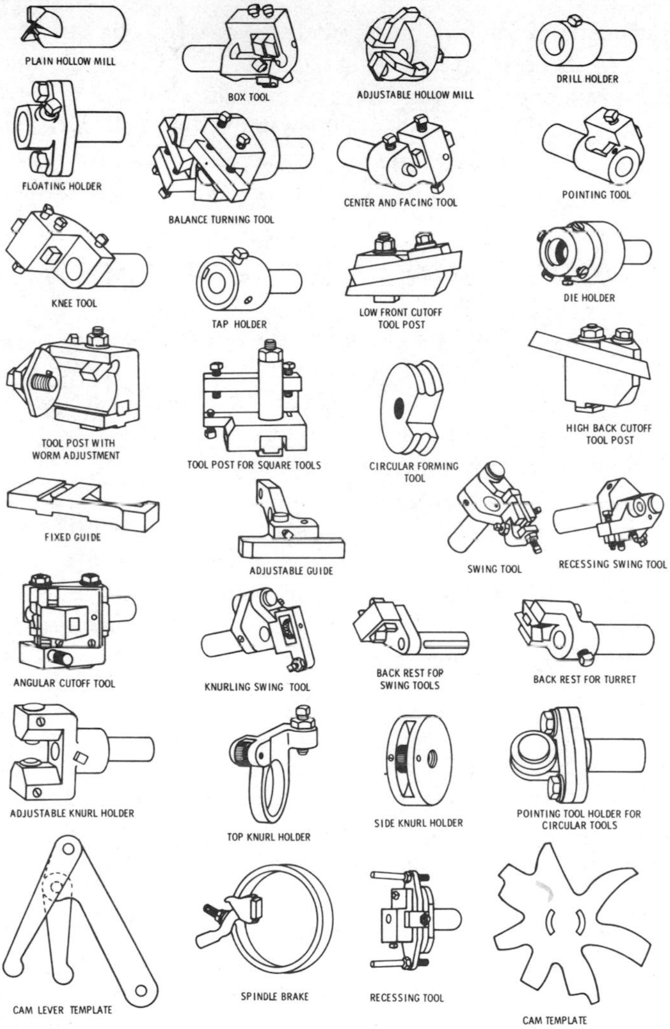

Fig. 13-4. Automatic screw machine tools and accessories.

Knurling tools—Top and side knurl holders are used on the cross slides. Adjustable knurl holders and knurling swing tools are used for knurling from the turret.

Forming and cutoff tools—Circular forming tools and cutoff tools are used on the regular cross-slide tool posts with a worm adjustment. Square tools and thin straight blade cutoff tools are used on the cross slides. The angular cutoff tools are used in the turret.

Supporting and auxiliary tools—Swing tools, fixed and adjustable guides for operating swing tools, back rests for turrets, and spindle brakes are other equipment that can be used on automatic screw machines.

General Suggestions for Tool Selection

The balance turning tool is usually for straight rough turning with plain hollow mills and adjustable hollow mills as the next choices. The box tools are preferable for straight finish turning; but adjustable hollow mills are sometimes used. The knee tool is used for turning scale on machine steel. A pointing tool of the box type is preferred for straight turning on long and slender work, as the bushing ahead of the blade holds the work steady.

The swing tool is used for taper turning and for some form turning where the curve is long, shallow, and continuous. Circular form tools are used on the cross slides for all form turning and for straight turning behind shoulders. Generally, they are used wherever they can be applied, and the length of cut is not great in proportion to the diameter. Extra-wide circular form tools are held in the tool post with worm adjustment. The swing tool can be used if the turning cannot be accomplished otherwise. Form turning can be performed occasionally with a straight-bladed tool held in the tool post for square tools. However, this is seldom used as the tool has no advantage over the circular form tool, and it does not last as well as the circular tool.

The center drill clamped in a floating holder is usually used for centering the end to be drilled, as this setup permits easy adjustment. If the end of a bar requires facing, as well as centering, a centering and facing tool that combines the two operations can be used.

To round off burr, or point the end of work, a pointing tool holder with a circular tool should be used if the work is strong enough to prevent springing during the operation. If the work is long and slender, use a pointing tool of the box type, which supports the work just ahead of the cut.

Drills, reamers, and counterbores should be mounted in a floating holder so that they can be easily adjusted centrally with the work. A drill holder is not used, unless the tool is so fine that it will spring enough to correct any inaccuracy in setting. A groove or recess inside the piece can be cut with the recessing swing tool.

Taps and dies should be held in nonreleasing tap and die holders. The releasing type should be used only in instances where it is not possible to develop a thread lobe on the cam because of its slow movement, and the tool must dwell before backing out.

If a thread is to be rolled, the top knurl holder for the cross slide is often used for carrying the thread roll; the cutoff operation immediately follows the threading operation on the same tool post. If the thread rolling operation is heavy, the side knurl holder is preferred for the cross slide, as it is free from any tendency to pull up on the cross slide. If there is no room on the cross-slide posts for the holders to carry a thread roll or if the thread is to be placed in such a position that tools on the cross slide cannot be used, a thread roll in a swing tool can sometimes be used. Frequently, it is advantageous to roll a short thread on brass, aluminum, or other soft metals where the thread cannot be cut with a die. The opening die holder is used in place of the button die when the spindle is run in only one direction.

The adjustable knurl holder is preferred for knurling on straight work because it carries two knurls that swivel for straight or diamond knurling and give a balanced cut. If the part to be knurled is not straight, but is formed—convex for example—the top or side knurl holders are used; this also applies if the knurling is behind a shoulder. The knurling swing tool is applied only on formed knurling or behind shoulders when the cross-slide holder is not available. The top knurl holder is usually preferable to the side knurl holder because it permits a cutoff tool to be used on the same tool post with it.

For straight cutoff work where burring, rounding, or forming of the bar or piece cut off is not required, a thin cutoff blade of the straight type held in a cutoff tool post is used. The circular cutoff tool is used most frequently because it can be shaped readily to part the piece properly from the bar, performing a chamfering or similar operation at the same time; once formed, the work retains its shape after repeated grindings.

If the cutoff or back end of the work is to be pointed or coned, the angular cutoff tool should be used. The fixed guide is intended only as a pusher for forcing the angular cutoff tools, as well as the various swing tools, when they are not fed along the piece against the work. If a swing tool is to be fed along the piece for turning after being forced into position by the guide, the adjustable guide should be used. This also applies to straight, taper, or formed work because adjustment of the guide can compensate for any slight error in the tool or machine.

In certain instances where a cut requires considerable side pressure that tends to spring the work, as in thread rolling, knurling, or forming operations, a back rest should be used in the turret for support if the work permits. Swing tool cuts on slender work occasionally require the use of a back rest in the swing tool shank. Back rests should not be used unless they are really necessary.

HOW TO SET UP AN AUTOMATIC SCREW MACHINE

The tooling and setup preparations given here are relative to the Brown & Sharpe No. 2 automatic (high speed) screw machine. Instructions given here are for the operator who has been furnished the cams, necessary tools, and a drawing. The method of setup is practically the same as for other machines. Some types are equipped with a speed-change mechanism that requires adjusting an additional set of dogs, but the general outline of operations is the same. Ingenuity and mastery of the technique are essential in tool selection and setup to realize the greatest measure of economy and efficiency that are possible on high-production automatic screw machines.

The general sequence can be followed in practically all instances, but familiarity can develop slightly different ways of making the various settings and adjustments. It will be noted that the tools are set up and adjusted in the same general order of their operations as that laid out for the job. For the particular job to be used here as an example, the data regarding dimensions of the piece, outlines of the cams, speeds, etc., are shown in Fig. 13-5. The outlines of the three cams are superimposed on each other; draw-

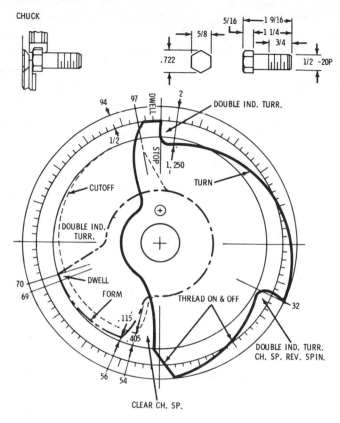

Fig. 13-5. Drawing of the piece and cams with the order of operations. The job data are given as follows: double-index turret; No. 22F left-hand box tool; double-index turret and reverse spindle; thread off; clear and double-index turret; form—back slide; cutoff—front slide; feed stock to stop (4 inches). Spindle speed: 165 r/min forward; 765 r/min reverse. Driving shaft speed, 240 r/min.

ing the cams in this manner makes the sequence or order of operations more clear at a glance. It is a good practice to list the order of operations with the number of spindle revolutions and feeds required for each operation, and the kind and size of tools that are to be used. If the machine has been used previously for a different job, it is necessary to remove the tools and cams and arrange the machine for either single or double indexing (Fig. 13-6).

274

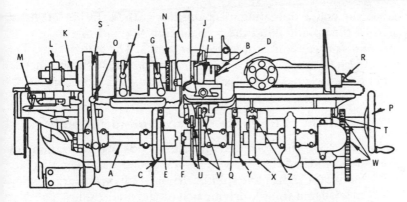

Fig. 13-6. Diagram of Brown & Sharpe No. 2 automatic screw machine.

The machine is set with the chuck open so that the grip of the new collet can be adjusted and checked as follows:

1. Raise the trip lever Q (See Fig. 13-6).
2. Turn the handwheel P until the chuck is open; lift latch L, and pull out the feed tube K.
3. Take out the feeding finger, which is threaded into the tube (left-hand thread), and replace with the feeding finger to be used on the new job.
4. Replace the feed tube, but do not leave the latch L down.
5. Unscrew the chuck nut on the end of the spindle with a pin wrench, and the spring collet will slip out.
6. Do not disturb nut J, as it is used to adjust the spindle box.
7. Install a new spring chuck, and before turning on the chuck nut, make sure that the sleeve into which the collet is inserted is completely back in place. This sleeve, like the collet, is free when the chuck nut is removed and is sometimes partly pulled out with the collet.
8. There is a slot in the back end of the sleeve that fits over a pin inside the spindle; turn the chuck nut tightly against the spindle nose.

The grip of the spinning chuck is adjusted by opening and closing the chuck with a hand lever inserted in the chuck form at G. The grip of the collet on the stock is regulated by means of the knurled nut N (see Fig. 13-6). On the No. 00 and No. 0 machine sizes, the chuck operating mechanism is at the extreme left-hand end of the spindle, and the two nuts on the end of the spindle provide for regulating the grip of the chuck. The grip of the collet should be firm enough that it will not let the work slip but not so tight that the levers that operate the chuck will be broken.

Arrangement of Belts for Correct Spindle Speed

In Fig. 13-5, the specified spindle speeds are 165 r/min for forward speed and 765 r/min for reverse speed. The diagram (Fig. 13-7) shows the right-hand spindle driving belt on the large pulley and the left-hand spindle driving belt on the small pulley. The first countershaft should run at 215 r/min with the belt on the third cone pulley.

Slow forward speed for threading on, and fast reverse speed for threading off, should be arranged for threading jobs requiring

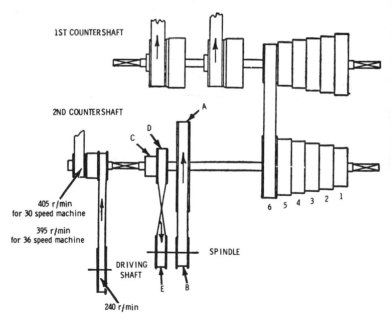

Fig. 13-7. Belt diagram for Brown & Sharpe No. 2 automatic screw machine (high speed).

reversal of the spindle. This same reverse speed is used for cutting operations using left-hand tools.

The spindle speeds are changed on motor-driven machines by means of change gears and spindle pulleys (sprocket on No. 2G machine) located in the base of the machine. A switch is located at the right-hand end of the tank table for reversing the motor as required. After reversing the motor, make certain that the back-shaft driving belt is running in the correct direction.

Indexing the Turret

Indexing is performed by a simple Geneva movement, and the turret is locked by means of a taper bolt that is withdrawn during the index movement by a cam connected with the turret change mechanism. There are two kinds of indexing: (1) single and (2) double.

The diagrams in Fig. 13-8 show the position of turret change rolls and turret locking pin cam for both single and double index-

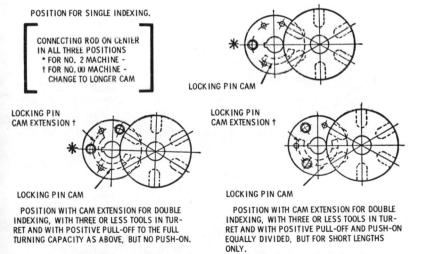

POSITION FOR SINGLE INDEXING.

CONNECTING ROD ON CENTER
IN ALL THREE POSITIONS
* FOR NO. 2 MACHINE -
† FOR NO. 00 MACHINE -
CHANGE TO LONGER CAM

LOCKING PIN CAM

LOCKING PIN
CAM EXTENSION †

LOCKING PIN CAM

LOCKING PIN
CAM EXTENSION †

LOCKING PIN CAM

POSITION WITH CAM EXTENSION FOR DOUBLE INDEXING, WITH THREE OR LESS TOOLS IN TURRET AND WITH POSITIVE PULL-OFF TO THE FULL TURNING CAPACITY AS ABOVE, BUT NO PUSH-ON.

POSITION WITH CAM EXTENSION FOR DOUBLE INDEXING, WITH THREE OR LESS TOOLS IN TURRET AND WITH POSITIVE PULL-OFF AND PUSH-ON EQUALLY DIVIDED, BUT FOR SHORT LENGTHS ONLY.

(A) Position for single indexing.

(B) Position for double indexing.

Fig. 13-8. Diagram showing positions of turret change rolls and turret locking pin cam for single and double indexing.

ing. One roll on the turret change gear, with the proper cam for withdrawing the turret locking pin, is used for single indexing. Two rolls (using only alternate holes in the turret), with an additional cam surface to withhold the turret locking pin the required amount of time, are used for double indexing.

Changing from Double to Single Index

The method of changing from a double to a single index is similar on all three sizes of machines except that the turret locking pin cams are in slightly different positions. Remove the cover from the turret slide, and remove the turret change roll. This is the roll not aligned with the eccentric pin on the turret change shaft. It is mounted on a stud held in place by a wedge pin which can be driven out with a steel punch and hammer. The procedure is as follows:

1. Remove the section of the cam that is square and without rise, as this is the part of the cam that holds the turret locking pin out of the turret during indexing of the second hole.
2. Trip the lever, and turn the handwheel to the completion of one index. Always try the mechanism by hand to avoid breaking any parts if the change has not been made correctly.
3. Replace the turret slide cover and proceed with the setup.

The change gears are replaced by taking off the change gears at W (see Fig. 13-6), which were used for the previous job, and replacing them with the gears required for the new job. These gears are specified on the drawing (see Fig. 13-5). Then arrange as shown in Fig. 13-9.

Setting Cross-Slide Tools

When the top of the work revolves toward the cross-slide post, a rising block D (Fig. 13-10) is used under the post to bring the tool to the proper height. Loosely clamp the circular cutoff tool to the proper cross-slide tool post by the bolt B, and hook bolt A, setting the cutting edge as close as possible to the height of the center of the work.

The cutting edge of the circular tool is adjusted to the work by turning nut E which, by means of an eccentric, tilts the thin plate F

CAM DESIGN WORK SHEET

PART NAME OR NO...... Example 1 ....MATERIAL ......BRASS ....
SURFACE FEET FOR STOCK............
            "   "   "   THREAD ....
            "   "   "   DRILL..............
            "   "   "   TURN..............

SPINDLE { FORWARD...1200 ....
R P M { BACKWARD ....
SECONDS...30 ....
GROSS PRODUCTION PER HOUR...120....
MADE ON...No. 2 ....

| Order of Operations | Throw | Feed | Spindle Revolutions | | | | | Hundredths | |
			For each Operation	After Deducting for Operations Overlapped	Readjusted to Equal Revolutions Obtainable with Regular Change Gears	For Spindle Revolutions in Preceding Column		For Operations That are Overlapped
Feed Stock to Stop				24		24	4	
Index Turret				24		24	4	
Rough Turn - No. 22 Balance Turning Tool.........	1.500	.010	150	150		150	25	
Index Turret.................				24		24	4	
Finish Turn - No. 22B Box Tool	1.500	.010	150	150		150	25	
Clear............................				24		24	4	3
Form - Front Slide ....{	.102	.003	60					10
	.010	.001	18					3
Index Turret 4 Times								16
Cut Off - Back Slide	.574	.0033	175	175		204	34	
96 Hundredths equal..					.547			
Estimated Total Revs., if spindle runs 1200 R P M continually....................					570			
Nearest Actual Spindle Rev. available on Machine						600	100	

NOTE: A dimensioned pencil sketch of the piece is often
drawn in blank space at top of this sheet.

Fig. 13-9. Cam design work sheet.

to which the tool is clamped. The clamp screw *K* on the post of the No. 2 size of machine keeps the eccentric from turning and should be loosened before adjustment. When the tool is set correctly, clamp it securely by means of bolts *A* and *B*, as shown in Fig. 13-10. (Also see Fig. 13-11.)

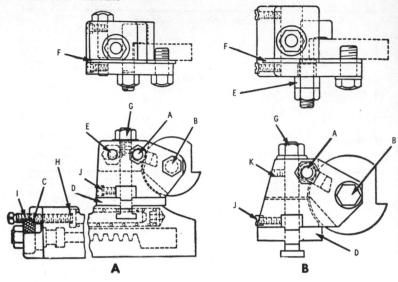

Fig. 13-10. Clamping and adjustment of cross slide circular tools.

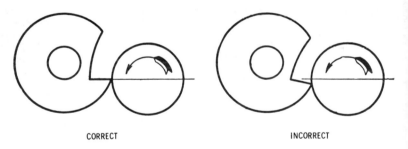

CORRECT   INCORRECT

Fig. 13-11. Correct (left) and incorrect (right) methods of setting the circular tool with reference to the work.

Adjusting the Cutting Tool to Proper Distance from Chuck

It is the most desirable practice to work as close to the chuck as possible with all tools. The cutoff tool is the one that governs the distance from the chuck more often than the others. The cutoff tool is set by loosening the nut G (see Fig. 13-10) and sliding the tool post either toward or away from the face of the chuck. In some instances it is the form tool, and sometimes it is a turret tool, that governs the distance to set the cutoff tool from the chuck face, but a study of each job and the tools to be used will enable the operator to determine the proper distance.

Adjust the Form Tool to Line Up with the Cutoff Tool

The stock should be fed out far enough for the cutoff tool to cut off a thin piece of stock. Start the spindle, place the hand lever in the hole in the cam lever that operates the cross slide on which the cutoff tool is mounted, and raise the lever until the tool slightly nicks the bar. Then stop the spindle and bring the form tool up in the same manner; align it with the groove cut by the cutoff tool, according to the way in which the job was laid out.

Placing the Cams

Preliminary to this operation, draw the cross slides back by means of the nut C (see Fig. 13-6). Care should be taken not to place the cams on backward. The upper side of the camshafts are "coming."

Adjusting the Cutoff Tool to the Cam Lobe

Turn the handwheel P (see Fig. 13-6) until the cam lever operating the cross slide that carries the cutoff tool is at the highest point of the cam lobe and about to drop back. Turn the nut C (see Fig 13-10) with a wrench, feeding the cutoff tool inward until the piece is cut off and the tool has traveled far enough past center to leave the end of the bar smooth. Turn the handwheel P (see Fig. 13-6) until the roll drops off the high point of the cross slide cam and the cutoff tool withdraws until it clears the diameter of the stock.

Adjusting the Turret to the Correct Distance from the Chuck

With the lead lever roll on top of the cam stop lobe, measure with a rule from the face of the chuck to the turret, and adjust the turret slide by means of a screw R (see Fig. 13-6) until the turret is at the required distance specified on the cam drawing. On cams designed for operating the swing stop, the turret adjustment figure is given from the top of another convenient lobe.

Setting the Stock for Length

Without changing the setting of the machine, place the stock stop in one of the turret holes, hold out the turret locking pin, and revolve the turret until the stop is in line with the spindle. Move the cutoff tool toward the center of the work with the hand lever of the cross slide, and then measure from the front edge of the cutoff tool a distance equal to the length of the piece. Set the stock stop to this point, clamp it securely (see Fig. 13-6), and turn the handwheel until the swing stop has reached full downward position. Then adjust the swing stop arm to the center of the stock and to the proper distance from the cutoff tool.

Setting the Chuck and Feed Trip Dog

The dog is fastened to the left-hand side of the carrier Y (see Fig. 13-6), and operates the trip lever Q. The dog should be set to trip the lever just as the cutoff tool clears the diameter of the stock. Sometimes, in close timing of jobs, the lever is tripped just before the cutoff tool clears the diameter of the stock. This is possible because the chuck has to open before the stock feeds; and by the time the chuck is open, the cutoff tool is out of the way.

Setting Turret Indexing Trip Dogs

Turn the handwheel P (see Fig. 13-6) until the lead lever roll T begins to drop off the stop lobe of the cam. Then set one of the trip dogs on either side of the carrier X so that it trips the lever Z and sets the indexing mechanism in operation. Turn the handwheel, and observe that the indexing of the turret consists of three movements: (1) the turret slide withdraws; (2) the turret then revolves from one station to the next; and (3) the turret slide advances into position again, where its farther forward movement is controlled by the outline of the lead cam.

When the turret slide has advanced into position, the lead roll lever should be at the bottom of its drop on the cam, as shown at A (Fig. 13-12). If the roll has dropped only a portion of the distance, as at *B*, the following turret must be "jabbed" into the end of the work when the slide advances into position. This means that the trip dog should be moved to trip the indexing mechanism slightly later. The easiest way to tell when the roll begins to drop off a cam lobe is to feel the motion of the turret slide by pressing the thumb against the back of the slide and its bearing on the machine bed, while turning the handwheel.

After setting the trip dog for the first index point, turn the handwheel until the roll begins to drop off the high point of the next lobe, and set the indexing trip dogs until the index point just prior to the thread lobe is reached. When setting the dog at this point, the spindle reverse trip dog should also be set. (See Fig. 13-13.)

Setting the Spindle Reverse Trip Dog

Turn the handwheel until the roll just begins to drop off the high point of the thread lobe, which can be felt at the joint of the turret slide and machine bed as described previously. At this point, set the trip dog on the side of the carrier *C* (see Fig. 13-6) to operate the trip lever *E* and reverse the spindle.

Setting the Indexing Trip Dogs

If threading is the last turret operation, set the necessary number of trip dogs required to rotate the turret so that the stock stop is in position when the stop lobe on the cam comes from under the lever roll. If only three tools are to be used, change to double indexing, and place the tools in alternate turret holes. Use three dogs on the turret dog carrier.

Adjusting the Feed Slide for Length of Stock

The crank *M* on the feed slide (see Fig. 13-6) provides a means of changing the length of stock fed whenever the stock feeding mechanism is in operation, and a scale on the feed slide bracket shows the approximate length fed. The feed slide should be set to allow not more than ⅜ inch in excess of the desired length. The stock should be set to feed out to the stock stop, which has already

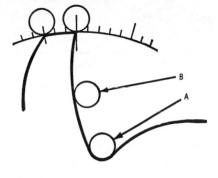

Fig. 13-12. Diagram showing position of lead lever roll at beginning (A) and at end (B) of indexing.

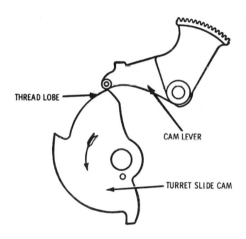

THREAD LOBE

CAM LEVER

TURRET SLIDE CAM

Fig. 13-13. Position of lead lever roll on thread lobe when the spindle reverses.

been set. After setting the feed slide, throw in the latch *L* (see Fig. 13-6), start the spindle, and throw in the starting lever *O*.

Placing and Adjusting the First Turret Tool

Throw out the lever *O* (see Fig. 13-6) when the roll drops off the first stop lobe, and let the spindle continue to run. Place the first tool in the turret hole next to the stop; set it well back into the turret, and clamp it securely. Place the hand lever for operating the turret in position, and feed the tool a short distance onto the work. The spindle can then be stopped, the turned diameter measured, and any adjustment that is necessary made on the tool to obtain the correct size.

Then, throw in the starting lever, or turn the handwheel until the roll is at the high point of the lobe, and stop at that point. The tool is fed onto the work by means of the hand lever, continuing until the correct length is turned. Adjustment should be made, loosening the bolt that clamps the tool in the turret and adjusting the tool outward to the proper length. Set the other tools by the same method, except threading tools, which should be left until the other tools have been set and the correct sample piece is obtained.

Adjusting the Form Tool

This tool is the last to work on the piece prior to cutting off; therefore, it is the last tool to be adjusted. On jobs where the forming operation takes place earlier, the tool should be adjusted as soon as the tools preceding it have been set.

The form tool should be adjusted with the lever roll on the highest point or dwell of the cross slide cam. The form tool can be adjusted to cut to proper depth by means of the nut C (see Fig. 13-10); the nut should be clamped by tightening the setscrew as soon as the correct diameter on the piece is obtained.

The positive stop H is provided to make certain that pieces are formed to accurate diameters. The most practical method of setting the stop is to advance the form tool by means of the nut C, until a sample approximately 0.008 inch under size is produced. With the cam roll on the full height of the cam, adjust the screw I until it holds the form tool back far enough to produce a sample piece of the correct size.

If the tool wears slightly, loosening the screw I slightly can correct the error; take care to set the tool to form parallel work. The two adjusting screws J, which shift the post with reference to its tongue, can correct the error. Before adjusting these screws, the nut G must be loosened slightly.

Then the machine is ready to start. First make a blank, and measure it carefully to see that all the tools have been set correctly.

Adjusting the Threading Tool

After the machine has been set to produce a suitable blank, it should be stopped just as the cam roll is positioned at the beginning of the thread lobe. Place the tap, or die, well back into the turret hole so that it will go only a short distance onto the piece.

If a solid die is used, make certain that the die is centered. If an opening die is used, the spindle pulleys can be run in the same direction; the slow speed can be used for threading, and the fast speed can be used for turning, drilling, etc. It is also necessary to use a die closing arrangement when an opening die is used.

In starting the machine, make certain that the spindle is running in the correct direction and at the correct speed. Start the machine, and make certain that the spindle reverse trip dog has been set correctly to operate just as the cam roll drops off the high point of the thread lobe. With the tap, or die, placed a short distance outward, make a sample piece. Check the dimensions, and continue to adjust until the correct length of thread is obtained.

Setting the Deflector

Clamp a dog in position on the right-hand side of the carrier Y (see Fig. 13-6). On light pieces and work that can be machined quickly, the deflector must remain under the chute a greater portion of one revolution of the cam shaft than on heavier pieces that drop instantly or on work that is machined slowly.

Setting the Automatic Stock

Adjust the spring in the tube until, with no stock in the spindle, its tension is great enough to throw the feed slide backward, thereby moving the roll on the feed operating lever from one side of the feed cam groove to the other. The machine will not operate if the tension is enough to overbalance the drag of the feeding fingers.

Measuring the Work

During operation of the machine, the pieces should be checked frequently. When working within fine limits, the work should be measured as each new bar is placed in the machine, as a variation in stock can affect the cutting action of the tools.

Renewing Stock

When the feed is stopped automatically at the end of a bar, the chuck is left open. Before inserting a new bar, the turret locking pin should be withdrawn by the thumb latch; and the turret should be turned to a point that is one-half the distance between stations, so that the short piece of stock left in the chuck can be pushed out by the new bar. Make certain that all burrs are removed from the ends of the new bar; turn the handwheel so that the chuck is open, and push the bar through until the end is just past the cutoff tool.

If the swing stop is being used, raise the stop by hand to allow the short end of the bar to drop out while inserting the new bar. Turn the turret back into position, and start the machine. The work should always be checked after insertion of a new bar.

DIAL-CONTROLLED MACHINES

The Cleveland Dialmatic machine has 112 spindle speeds, ranging from 24 r/min to 1820 r/min, and dial-controlled feeds. A two-speed motor supplies power for the spindle and provides two automatic spindle speeds. Two additional speeds are made available by an automatically shifted disk-type friction clutch. This makes a total of four automatically changed spindle speeds, both forward and reverse, that are available for each set of change gears.

Tool feeds are dial controlled (Fig. 13-14). An electric feed drive provides separate adjustable feeds for all turret positions. Feeds can be set quickly in these machines as follows:

1. Cam changes are not necessary to adjust feeds.
2. Spindle speeds and feeds can be transposed into dial setting by means of conveniently located charts.
3. A turn of a knob sets a feed.
4. Feeds in inches per minute are accurately indicated on the monitor.

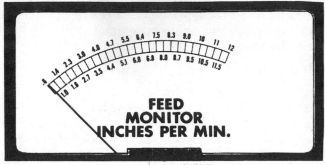

Fig. 13-14. The monitor permits accurate feed settings in inches per minute.

5. Ten different feeds are possible in a single cycle of the machine.
6. Feeds can be adjusted while the tools are cutting.
7. All feeds are infinitely variable.
8. A lock on the cover can protect dial settings after setup.

SUMMARY

The screw machine was originally designed to make small screws and studs. The flexibility of the machine has adapted it to a large variety of work.

A screw machine is a type of turret lathe designed to perform a series of operations at one time. The turret is provided with several tools, brought into a working position to perform various operations as the work is being turned.

Drilling, boring, reaming, forming, and facing are just a few examples of the operations that can be performed on a screw machine. The various types of screw machines can be classified as to the type of operation, such as plain or hand operated, semiautomatic or automatic, and to the type of spindle, such as single or multispindle.

Many different kinds of tools are used because of the wide variety of work performed on automatic screw machines. Turning tools, cutting tools, threading tools, knurling, forming, and cutoff tools are most generally used. Swing tools, fixed and adjustable guides for operating swing tools, back rests for turrets, and spindle brakes are other equipment that can be used on automatic screw machines.

REVIEW QUESTIONS

1. What are some advantages of the automatic screw machine?
2. Name the various types of screw machines.
3. What are some of the various operations performed on an automatic screw machine?
4. Name the different types of tools used on an automatic screw machine.

CHAPTER 14

The Milling Machine

A milling machine is a power-driven machine that cuts metal by means of a multitooth rotating cutter. The machine is constructed in such a manner that the workpiece is fed to a rotary cutter—instead of revolving as on a lathe, or reciprocating as on a planer.

ADAPTATION

A variety of operations can be performed on the milling machine. A multitooth milling cutter remains sharp much longer than a single cutting tool; and the cutting action of the milling machine is continuous, as compared to the intermittent cutting action of the shaper and planer. Therefore, many kinds of workpieces can be machined more economically on the milling machine than on a shaper or planer.

Milling machines are adapted for production of workpieces that have intricate profiles which must be interchangeable. Large castings or forgings can be handled on the milling machine. Many specialized types of milling machines have been developed, which has broadened considerably the range of application of the milling machine.

CONSTRUCTION AND CLASSIFICATION

The milling machine has a power-driven spindle. An arbor for holding multitooth cutters fits into the spindle. The cutting edges or teeth of revolving circular cutters remove a controlled amount of metal at each revolution of the cutter. The workpiece is mounted on a movable table and is fed against the cutter. The table can be moved either by hand feed or by power feed. When several cutters are mounted on the arbor, several surfaces can be machined in one operation. The knee-and-column type of milling machine is commonly found in industry (Fig. 14-1).

Knee-and-Column

The knee-and-column types are usually cast in one piece to make up the main casting of the milling machine. The column is thick walled and strongly braced to support the other parts of the machine, and usually includes a tank for coolant used on the cutter. The inner space of the column houses the driving motor and the gear mechanism for transmitting power to the spindle and table of the milling machine.

The top of the column supports the overarm or overarms. The design of the overarms depends on the manufacturer of the machine. The arbor support slides on the overarm.

The arbor support clamps onto the overarm and supports the milling arbor. The bearing in the lower end of the arbor support should be in perfect alignment with the spindle axis. The *inner support* supports the arbor near the middle; a bearing sleeve placed between the collars on the arbor fits into the large hole. A smaller hole in the *outer support* holds the ground end of the arbor.

The spindle is hollow throughout its entire length. It revolves in bearings in the upper end of the column; the front end has a

tapered hole for receiving the standard shanks of milling arbors
(Fig. 14-2).

On the face of the column, a wide slide, usually of dovetail
design, is provided to assure proper alignment of a sliding knee or
bracket. The knee can be raised and lowered on the column to

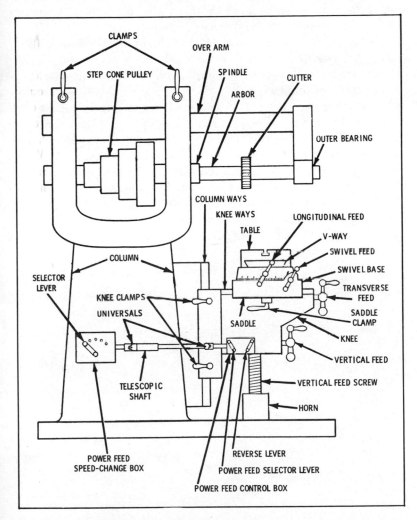

Fig. 14-1. Diagram of the universal milling machine, showing knee-and-
column design, basic parts, and controls.

adjust the depth of cut for jobs of various sizes. The knee is a casting with two of its sizes machined at right angles. The vertical machined side of the knee slides on the ways on the face of the column. The top of the knee is machined at a right angle to the vertical surface and supports the saddle, which slides on the knee either toward or away from the face of the column. A sturdy elevating screw raises and lowers the knee.

The saddle is mounted at the top of the knee and supports the table. A precisely machined surface, usually dovetails, on the top side of the saddle holds the milling machine table (Fig. 14-3).

The table is mounted on the saddle. The table movements of the *plain* milling machine are: (1) vertical—raising the knee on the column; (2) transverse—sliding the saddle on the knee; and (3) longitudinal—sliding the table on the saddle. Micrometer dials graduated in thousandths of an inch permit accurate setting or positioning of the table. The table can also be swiveled horizontally on the saddle of the *universal* milling machine.

The milling machine table has T-slots running lengthwise on its top surface. T-bolts can be used to fasten either the work or a work-holding device to the table.

Fig. 14-2. Spindle drive gearing of a milling machine.

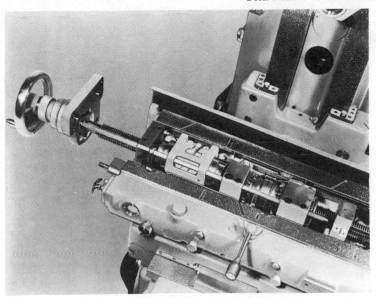

Courtesy Cincinnati Milacron Co.

Fig. 14-3. Saddle for a milling machine.

Types of Milling Machines

The knee-and-column type of milling machine is used more frequently in tool and die shops than any other kind of milling machine. The knee-and-column mill is classified as either horizontal or vertical, depending on the position of the spindle and table adjustments. The horizontal knee-and-column mill is divided into two classes: plain and universal.

Plain milling machine—The plain horizontal milling machine is very common in industry. The horizontal spindle projects at right angles from the column face. The spindle is hollow and tapered internally and permits the knee to be raised and lowered vertically on the accurately machined ways of the column. The table is mounted on machined ways on the saddle which rests on the knee; it can be moved either by hand or by power. The table feed (Fig. 14-4) on plain milling can be in three directions—longitudinal (at right angles to the column face), transversal (crosswise), and vertical.

295

Fig. 14-4. Feed unit for a milling machine.

Universal milling machines—The universal milling machine has the same table movements (longitudinal, crosswise, and vertical); and, in addition, the table can be set at an angle to the column face. The table is mounted on a swivel block which can be rotated about the center of the universal saddle. A modern universal milling machine is shown in Fig. 14-5.

Vertical milling machine—The spindle is in a vertical position on the vertical knee-and-column mills. The spindle can be raised and lowered, and the machine is especially adapted for boring and profile work. Since the cutting tool is held in the spindle, an arbor is not needed on the vertical mill. The table of the vertical milling machine can be used on longitudinal, transverse, and vertical feed movements. A vertical milling machine is shown in Fig. 14-6.

Special types of milling machines—In addition to the knee-and-column milling machines, a number of specialized types of milling machines are used for certain kinds of work. *Thread mills* are used to mill both external and internal threads. *Spline mills* are used to cut keyways and slots.

In *bed-type* milling machines, the table cannot be raised or lowered; however, the spindle can be raised or lowered. The table

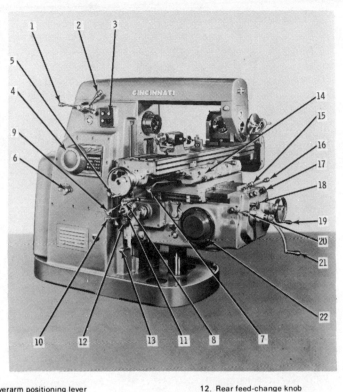

1. Overarm positioning lever
2. Overarm clamping lever
3. Rear push-button group
4. Spindle-speed dial
5. Table traverse handwheel
6. Spindle reverse lever
7. Rear table feed lever
8. Rear cross-feed lever
9. Rear rapid-travers lever
10. Knee clamping lever
11. Rear cross hand traverse

12. Rear feed-change knob
13. Rear vertical hand traverse
14. Automatic backlash eliminator knob
15. Rapid-traverse lever
16. Cross-feed lever
17. Front push-button
18. Front feed-change knob
19. Cross-traverse handwheel
20. Vertical-feed lever
21. Vertical hand traverse crank
22. Feed dial

Courtesy Cincinnati Milacron Co.

Fig. 14-5. Universal milling machine.

moves in a horizontal position with longitudinal feed only; it does not have a cross-feed movement. These are high-production machines.

The *duplex* milling machine has two opposed spindles. The spindles can be operated simultaneously, but each is separately driven and controlled. This type of machine has high productive

Fig. 14-6. Vertical milling machine.

capacity, as two cuts can be taken at the same time on similar or different workpieces. A roughing cut with one cutter and a finishing cut with the other can be taken when the machine has two spindles.

SUMMARY

The milling machine is a power-driven machine that cuts metal by using multitooth rotating cutters. The workpiece is fed to a rotary cutter instead of revolving as on a lathe or reciprocating as on a planer.

Many operations can be performed on the milling machine. The cutters remain sharper much longer than a single cutting tool, and the cutting action of the milling machine is continuous, as compared to the intermittent action of the shaper or planer. Large castings or forgings can be handled with ease on the milling machine.

The knee-and-column milling machine is used more frequently in tool and die shops than any other type. This machine is classified as either horizontal or vertical, depending on the position of the spindle and table adjustments. Knee-and-column milling machines are divided into two classes—plain and universal. There are also special types of milling machines, such as thread mills, used to mill both external and internal threads. A spline mill is used to cut keyways and slots. A duplex milling machine has two opposed spindles that can perform two separate operations at the same time.

REVIEW QUESTIONS

1. What are the advantages in using a duplex milling machine?
2. Which milling machine is most common in tool and die shops?
3. What is the basic design of a milling machine?
4. Into what two classes are knee-and-column machines divided?

CHAPTER 15

Milling Machine Operations

The operator of a milling machine, or any other machine, should be aware of certain safety precautions. A milling machine is, of course, not a plaything; but if proper precautions are taken, danger can be avoided.

SAFETY PRECAUTIONS

First, the operator should understand thoroughly the construction and operating action of the machine before attempting its operation. This is important not only for the operation but also because it lessens the chances of the machine itself being damaged seriously.

The following suggestions by Brown and Sharpe should be carefully noted and remembered:

1. Do not move an operating lever without knowing in advance the action that is going to take place.
2. Never toy with the control levers or carelessly turn the handles of the milling machine while it is stopped.
3. Do not lean against, or rest the hands on, a moving table; if it is necessary to touch a moving member, make sure of the direction in which it is moving.
4. Do not attempt to cut without being sure that the work is held securely in the vise or fixture and that the holding member is fastened to the machine table.
5. Remove chips with a brush or other suitable means—never with the fingers or hands.
6. Before operating a milling machine, study it thoroughly; then, if an emergency should arise, the machine can be stopped immediately.
7. Stay clear of the milling cutters; do not touch a cutter, even while it is not moving, unless there is good reason to do so, and then be careful.

The above suggestions, in general, can be applied to any other machine in the machine shop.

PRELIMINARY OPERATIONS

The various duties of the milling machine operator can be classified as: (1) preliminary operations and (2) machining operations. Several preliminary operations are necessary before the machine can be started.

Cleaning

Cleaning of the milling machine cannot be emphasized too strongly. The machine should be cleaned both before and after using. Accuracy and durability of the machine depend on its being kept clean—this applies to all machine tools.

If a milling machine is cleaned as soon as possible after use, it can be kept clean more easily than if chips and coolant are allowed to accumulate for long periods of time; a clean machine is usually in better mechanical condition than one in which dirt has been allowed to accumulate. This is because chips that build up on a dirty machine can clog oil channels, mar the bearing surfaces of the column or knee, and interfere with accurate adjustments of the fixtures and work. Oilholes that have become clogged with gummy oil can be flushed with gasoline—this will not injure the bearings.

Oiling the Machine

Apply a few drops of oil to each bearing at frequent intervals, rather than a flood of lubricant at long or irregular intervals; most bearings can hold only a few drops, and an excess of oil runs out and is lost. It is always safest to purchase a lubricant that is reliable, rather than experiment with less expensive machinery oils. It is less expensive to use a good oil than to risk damage to bearings from overheating or scoring.

The more recently designed milling machines are equipped with automatic lubricating systems that ensure a constant supply of clean lubricants at important points. This relieves the machine operator of much of his oiling problem, but the oil level in machine reservoirs should be checked at regular intervals.

Many machines equipped with automatic lubrication include oil filters in the system; all oil must pass through the filters each time it is circulated. The filters should be checked regularly to determine whether they are functioning properly; when the filter becomes clogged, the bearings can become dry, even though the oil is bypassed around the filter.

Mounting the Workpiece

Methods of mounting the work on the milling machine are similar to those for the shaper, planer, and other machines with worktables. However, work should be mounted more securely on the milling machine tables because the greater thrust produced by the cutter, especially the wider cutters, can damage either the work or the machine. Various methods of mounting the work on the milling machine for machining are: on the table; in a chuck; on an arbor; between centers; and on fixtures.

Clamping devices, such as T-bolts, toe dogs, pins, and screw pins, are similar to those used to hold planer workpieces (Fig.

15-1). The application of these devices is the same as for other machines, except that even more attention is required for supporting and anchoring the work to prevent springing. Stop pins should be used wherever possible to prevent movement of the work.

Much of the work that is performed on milling machines is mounted on fixtures. The table, fixtures, and all devices involved

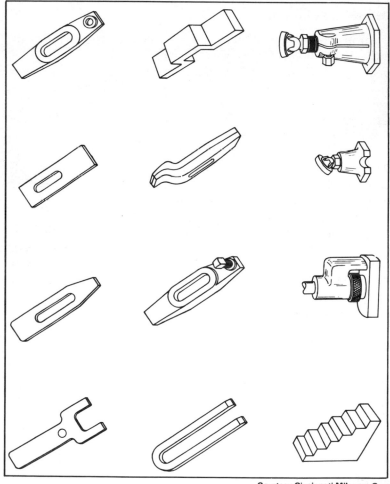

Fig. 15-1. Types of clamps, step blocks, and jacks commonly used in milling machine setups.

in mounting the work should be absolutely free from chips, dust, etc. Fixtures can be built to facilitate handling and to increase production if a large enough quantity of pieces warrants the expense.

For single-piece production or small-lot production, the work must be held by individual setup because it is impractical to construct special fixtures. The work must be positioned and held securely; the clamps must be carefully arranged, so that they will not be struck by the cutter or damage other parts of the machine.

The milling machine vise is the simplest holding arrangement. A special base for bolting to the table and a swiveling body permit adjustment of the vise to any angle in a horizontal plane. A scale, graduated in degrees, on the base of the vise is accurate enough for most work; but if accurate angles are required, the setting should be checked with a protractor. The toolmaker's universal vise can be used for work involving compound angles (Fig. 15-2).

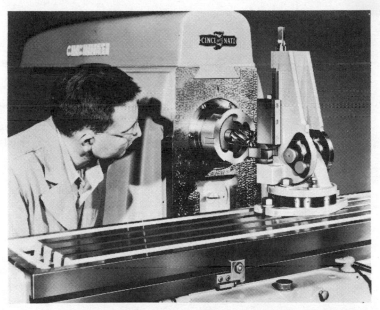

Courtesy Cincinnati Milacron Co.

Fig. 15-2. The toolmaker's universal vise can be used for general toolroom work. It swivels 90° in the vertical plane and 360° in the horizontal plane.

It is designed for adjustment in both the vertical and horizontal planes.

The standard vise is opened and closed by a crank. Hammering the handle of the vise to obtain a sure grip is often frowned upon because it can crack or break the screw or other parts, resulting in a serious accident when pressure is applied. However, many practical machinists insist upon the practice, maintaining that the work must be held securely. If the practice is followed, use a lead or Babbitt hammer for a single sharp blow, after tightening by hand as much as possible. A machinist's hammer should never be used on the handle of the vise.

The base of the vise should be cleaned carefully before installing it on the machine; then a dial indicator can be used to check for accurate alignment (Fig. 15-3). Clamp the dial indicator onto

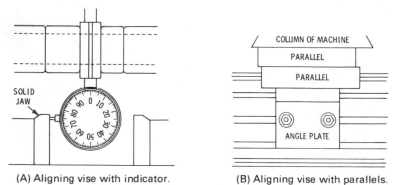

(A) Aligning vise with indicator.   (B) Aligning vise with parallels.

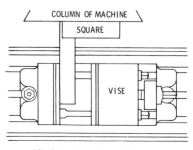

(C) Aligning vise with square.

Courtesy Cincinnati Milacron Co.

Fig. 15-3. Methods of aligning the vise jaws with the columns or spindle. The dial indicator method is the most accurate.

306

the machine arbor; then move the table so that the indicator rests on the fixed, not the adjustable, jaw. Set the indicator dial to zero, and by means of the manual feed, traverse the jaw past the indicator. Adjust the vise until the indicator shows no variation between the two ends of the jaws.

Another method of aligning the vise is to place the stock of a machinist's square against the face of the machine column and bring the blade into contact with the fixed jaw—with tissue paper feelers between the blade and the jaw. If an extra pull is required to remove all feelers, the vise is aligned (see Fig. 15-3).

Selecting the Cutter

Failure to obtain satisfactory results on a job can often be attributed to improper selection of the milling cutter; or, even if the proper cutter is used, working conditions can be unfavorable to proper performance of the cutter. Either the operator or another person in the milling department should be proficient in the use and care of milling cutters and capable of determining the correct feeds and speeds for their operation. A brief summary of the general application of various milling cutters is as follows:

1. Double-angle cutter—Used for fluting taps, reamers, etc.
2. End mill—Suited for surface milling, profiling, slotting, etc.
3. Fly cutter—Can be used to mill intricate shapes that do not warrant the expense of formed cutters.
4. Formed cutters—Used to cut curved or irregular surfaces; also for fluting taps, reamers, twist drills, etc.
5. Half side cutter—Used for face milling—or in pairs for "straddle milling."
6. Helical cutter—Especially adapted for thin work or intermittent cuts where the amount of slack to be removed varies. It is not a general purpose cutter. This cutter is erroneously called a "spiral" cutter.
7. Inserted-tooth cutter—gives long life; used for heavy duty face and side milling work.
8. Interlocking side cutters—Designed to maintain an exact width for milling slots.
9. Nicked cutters—Suited for making deep cuts as in roughing work.

10. Plain cutters—Used to mill a flat surface parallel to its axis.
11. Side cutter—Suitable for slotting operations.
12. Single-angle cutter—Chiefly used to cut milling machine cutter teeth.
13. Slitting cutter—Used to cut deep slots and to cut off stock.
14. Slotting cutter—Used to mill shallow slots, such as screw-head slots.

In addition to the above summary of cutter applications, the final selection of the proper milling cutter can be influenced by other factors as:

1. Milling cutters with comparatively few widely spaced teeth can remove more metal in a given time (for certain kinds of milling work) than cutters with a comparatively large number of teeth, without stressing the cutter or overloading the machine.
2. Fine-tooth cutters are inferior to coarse-tooth cutters, so far as their relative tendency to "chatter" is concerned.
3. The free cutting action of coarse-teeth cutters is due chiefly to the fact that less cutting action is required to remove a given amount of metal; each tooth takes a larger, deeper chip.
4. Wide spaces between the teeth allow the cutting edges to be well backed up, which is not always possible with teeth that are closely spaced.
5. Moderate rake angles reduce power consumption and are desirable on cutters that are used on mild steel; large rake angles are undesirable because of the tendency to chatter.
6. The helical angle has little effect on power consumption; however, a large helical angle is desirable because it requires fewer cutter teeth, gives smoother cutter action, and reduces the tendency to chatter.
7. With few exceptions, coarse-tooth cutters are superior to fine-tooth cutters for production work.

Sharpening Milling Cutters

Do not permit a cutter to become dull. Like the barber—sharpen the cutter often. Experience has shown that a dull cutter wears more rapidly than a sharp one and does not last as long as a cutter kept in proper condition.

The clearance or relief of a milling cutter is the amount of material removed from the top of the teeth, back of the cutting edge; this permits a tooth to clear the stock after the cutting edge has done its work. On formed cutters it is not necessary to consider clearance; this is because the teeth are formed so that the clearance angle remains the same when the face of a tooth is ground. Rake and clearance angles are illustrated in the profile of a milling cutter (Fig. 15-4). The narrow space or land A back of the cutting edge is about ¹⁄₃₂ inch in width and is ground to a clearance angle B. This clearance angle, together with the rake angle C on the front of the tooth, varies with the material being cut. Changing either the clearance angle or the rake angle (or both) only slightly, can make a vast difference in the operation of the milling cutter on the workpiece.

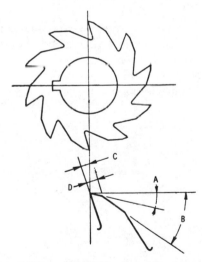

Fig. 15-4. Profile of a milling cutter showing: (A) clearance angles, (B) secondary clearance, (C) rake or undercut, and (D) land.

The clearance angles of milling cutters should always be checked. The fact that a milling cutter has a keen cutting edge does not necessarily mean that it will cut satisfactorily. As a general rule, the clearance angle for plain milling cutters over 3 inches in diameter should be 4°; the clearance angle for plain cutters less than 3 inches in diameter should be 6°. The Cincinnati Milacron Company recommends the following clearance angles:

Aluminum .... 10°
Bronze, cast .... 10-15°
Cast iron
  fast feeds...................................... 7°
  moderate feeds .... 6-7°
Copper .... 7-10°
Steel
  castings .... 6-7°
  hard (tool steel) .... 4-5°
  low-carbon .... 5-7°

The clearance angle for helical mills depends on the material being milled, depth of cut, etc. generally, a 5 to 7° clearance angle for cast iron and a 3 to 4° clearance angle for machine steel are satisfactory.

Details of a correctly sharpened side mill for milling cast iron are shown in Fig. 15-5. The illustration shows a 3/64 inch land that is ground at an angle of 6°, which is the clearance angle. Immediately behind the land, the tooth of the cutter is ground at an angle of about 12°. This angle should be no larger than is necessary to

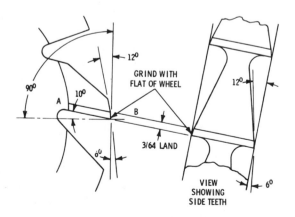

Fig. 15-5. Details of a correctly sharpened milling cutter.

prevent the heel of the cutter dragging on the work. The side teeth should be ground in exactly the same manner—a $\frac{3}{64}$ inch land, a 6° clearance angle, and backed off at a 12° angle. If the side mill were used on steel, the proper clearance angle would be 4°, rather than 6°. If this kind of cutter shows a tendency to chatter, the clearance angle on the side teeth should be reduced to as small as 1°. Conditions can be improved even more if the sides are slightly hollow ground, that is, if the face of the cutter is ground thinner at the inner end of the side teeth (A in Fig. 15-5) than at the outer end B. Fixed rules cannot be given for the clearance angles on cutters, as this depends on the material being machined, the depth of the cut, the style of the cutter, etc.

Mounting the Cutter and Arbor

To obtain rigidity in the setup, select a cutter that is large enough to prevent using an arbor of large diameter; this is necessary to obtain best results in heavy-duty milling. Place the cutter (or cutters) on the arbor as near the end of the spindle as the work permits to prevent springing the arbor (Fig. 15-6).

Courtesy Cincinnati Milacron Co.

Fig. 15-6. Typical milling machine arbors.

The support arm, which carries the outer bearing and the intermediate bearings, should always be used to provide proper support for the arbor (Fig. 15-7). The arbor support arm carrying the outer bearing should be in place before tightening the arbor nut to avoid danger of bending the arbor.

Milling cutters preferably should have a key drive rather than a friction drive. If a key drive is used, it is not necessary to hammer the arbor wrench to tighten the nut; it can be tightened sufficiently by hand. Various cutter mounting and arbor supports are illustrated in Fig. 15-8.

Fig. 15-7. An overarm provides rigid support for the arbor and cutters, maintains accurate alignment, and dampers vibrations set up by cutter action.

Short arbors (see Fig. 15-8A) are provided with a pilot bearing at the end of the arbor. This bearing fits a split bronze bushing X in the arbor support.

Some medium-length arbors (see Fig. 15-8B), in addition to an end pilot bearing X, have an arbor bearing collar to fit the intermediate support Y. This support should be placed as close to the milling cutter as is practical; the cutter should be located as close to the shoulder of the arbor as conditions permit. Another style of medium-length arbor (see Fig. 15-8C) does not have the pilot bearing for the bronze bushings at the end of the arbor, but it is provided with a bearing collar for placing the arbor support near the cutter.

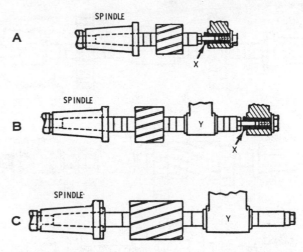

Fig. 15-8. Various cutter mountings and arbor supports for: short arbor (A); and medium-length arbors B and C.

Long arbors (Fig. 15-9A) should have two support bearings whenever possible; one support Y should be placed between cutters that are spaced at a distance on the arbor, and the other support Z, to which the braces are fastened, should be as near the outside cutter as conditions permit.

If the width of the table does not permit bringing the outer support Z, to which the braces are fastened, near the cutters, the intermediate support Y can be placed near the outer cutter, between it and the outer support (see Fig. 15-9B).

In some instances, the nature of the work requires the cutters to be placed near the outer end of the arbor (see Fig. 15-9C). Then the intermediate support Y should be placed between the inner cutter and the spindle.

An incorrect method of mounting a milling center on a long arbor is illustrated in Fig. 15-9D. This kind of setup should never be used, because it cannot possibly produce results that are satisfactory.

Direction of Cutter Rotation

The direction of cutter rotation in relationship to direction of feed is highly important. Many cutters can be reversed on the

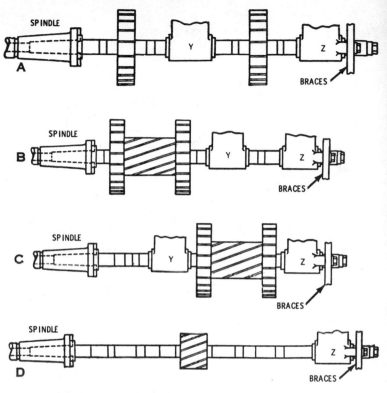

Fig. 15-9. Various cutter mountings and arbor supports for long arbors. Correct methods of mounting are shown in (A), (B), and (C); an incorrect method of mounting is shown in (D).

arbor, so it is important to know whether the spindle is to rotate clockwise or counterclockwise. In this connection it is also important to know the direction of feed. The direction of feed is opposite the direction of cutter rotation in *conventional milling* (Fig. 15-10). This can be called "up-cutting," which means that the cutting edge starts at the bottom of the cut and removes a progressively larger chip as the feed progresses. Formerly, all milling was done in this manner, but it has a disadvantage in that a certain amount of rubbing occurs before the tooth takes hold, and a series of small scallops shows on the finished surface.

The direction of feed is in the same direction as the direction of cutter rotation in *climb milling* (see Fig. 15-10). This action is also

314

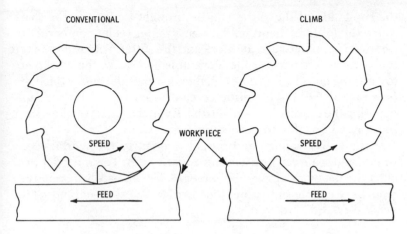

Fig. 15-10. Conventional milling action (left) and climb milling action (right).

known as "down milling" or "hook milling." Fast cutting rates and a better finish are obtained in climb milling. The pressure of the cut tends to hold the workpiece downward on the table. However, climb milling cannot be used on older machines unless they have a backlash eliminator on the feed screw nut—except where the work must be climb milled and light cuts can be taken. If climb milling is attempted on an unsuitable machine, the work can be dragged under the cutter, which will jam and break, and the machine can be seriously damaged. Most modern milling machines are designed for climb milling.

Locating the Cutter

For most milling operations, the position of the cutter in relationship to the work is not of great importance. Usually, the worktable can be moved up or down and forward or backward until the cutter barely touches the work. If it is necessary to position the cutter accurately in relationship to a flat surface, a square or straightedge can be used to align the cutter with the surface; then the work can be moved the necessary amount by means of the manual controls.

If a round surface is involved, such as milling a keyway in a shaft, a different positioning technique is required (Fig. 15-11). If the front side of the cutter can be brought tangent to the shaft, move the table by hand until a tissue paper feeler can barely be removed from between the work and the cutter. Set the cross-feed dial to zero. Then lower the worktable and move the workpiece toward the cutter a distance equal to one-half the cutter thickness (see Fig. 15-11A). This distance can be measured by carefully counting the graduations on the dial. Even the paper thickness can be allowed for, if extreme accuracy is desired.

If the cutter cannot be brought tangent to the workpiece, a square and scale can be used (See Fig. 15-11B). To set for depth of cut, raise the worktable until a paper feeler can barely be removed from between the cutter and the top of the workpiece, and set the vertical-feed dial to zero.

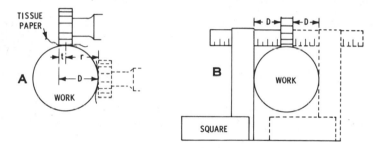

Fig. 15-11. Two methods of locating the cutter in relationship to the center of a round workpiece: (A) tissue paper method, and (B) use of square and scale.

SPEEDS AND FEEDS

Speed and feed rates are governed by several variable factors: material, cutter, width and depth of cut, required surface finish, machine rigidity and setup, power and speed available, and cutting fluid. Even though suggested rate tables are given, individual experience and judgment are extremely valuable in selecting correct milling speeds and feeds. The lower figure in the table for a given material should always be used until sufficient practical experience has been gained to change to the highest figure. Speed can be increased until either excessive cutter wear or chatter indicates that the practical limit has been exceeded.

Speeds

Unlike lathe tools, the cutting action for each tooth of a milling cutter is intermittent; there are relatively long cooling intervals, with the cooling intervals for large-diameter cutters being longer, of course, than for smaller cutters. The introduction of high-speed steel and the newer, faster-cutting alloys for construction of the cutters has considerably increased cutter speeds, as compared with the cutting speeds for the carbon steel cutters that were used formerly. Thus, the material used in construction of the cutters has an important bearing on permissible speeds (see Table 15-1).

Table 15-1. Cutting Speeds

(Surface feet per minute)

MATERIAL	High-Speed Steel		Carbide-Tipped		COOLANT
	Rough	Finish	Rough	Finish	
Cast Iron	50–60	80–110	180–200	350–400	Dry
Semi-steel	40–50	65–90	140–160	250–300	Dry
Malleable Iron	80–100	110–130	250–300	400–500	Soluble, Sulfurized or Mineral Oil
Cast Steel	45–50	70–90	150–180	200–250	Soluble, Sulfurized, Mineral or Mineral Lard Oil
Copper	100–150	150–200	600	1000	Soluble, Sulfurized or Mineral Lard Oil
Brass	200–300	200–300	600–1000	600–1000	Dry
Bronze	100–150	150–180	600	1000	Soluble, Sulfurized or Mineral Lard Oil
Aluminum	400	700			Soluble or Sulfurized Oil, Mineral Oil and Kerosene
Magnesium	600–800	1000–1500	1000–1500	1000–5000	Dry, Kerosene, Mineral Lard Oil
SAE Steels 1020 (coarse feed)	60–80	60–80	300	300	Soluble, Sulfurized, Mineral or Mineral Lard Oil
1020 (fine feed)	100–120	100–120	450	450	" " "
1035	75–90	90–120	250	250	" " "
X-1315	175–200	175–200	400–500	400–500	" " "
1050	60–80	100	200	200	" " "
2315	90–110	90–110	300	300	" " "
3150	50–60	70–90	200	200	" " "
4340	40–50	60–70	200	200	Sulfurized and Mineral Oils
Stainless Steel	100–120	100–120	240–300	240–300	" " "

Courtesy Cincinnati Milacron Co.

The amount of material to be removed per minute and the relationship between depth of cut and feed influence cutter speed. For example, a cut ⅛ inch in depth and ⅛-inch feed per revolution can be taken at a higher speed than a cut ¼ inch in depth and ¹⁄₁₆-inch feed per revolution, although the amount of material removed per minute would be the same in each operation. Cutter speed also can depend on the rigidity of the machine and the fixture in which the workpiece is held.

On some jobs, disregarding expense, the cutter can be run at excessive speed if it is reground frequently; or it can be run at a very slow speed and reground at longer intervals. Accordingly, speeds can be determined for a given cutter at which its greatest efficiency is achieved.

Cutter speeds are always given in surface feet per minute (s.ft./min), that is, the speed at which the circumference of the cutter passes over the work. Thus, the surface speed of a 6-inch cutter is twice as great as that of a 3-inch cutter for a given number of spindle revolutions. As spindle speed of a milling machine can be given only in revolutions per minute (r/min), it is necessary to translate, s.ft./min into r/min for making speed adjustments on the machine. If a table of speeds is not available, spindle speed can be calculated by the formula:

$$r/min = \frac{s.ft./min \times 12}{\times \text{ diameter}}$$

Likewise, s.ft./min can be calculated by the formula:

$$s.ft./min = \frac{\times \text{ diameter} \times r/min}{12}$$

Feeds

The feed rate is the rate at which the work advances past the cutter and is commonly given in inches per minute (i/min). Generally, the rule in production work is to use all the feed that the machine and the work can stand. However, it is a problem to determine where to start the feed. Table 15-2 and Table 15-3 give the suggested speed per tooth for high-speed steel and sintered carbide-tipped milling cutters respectively. Note that the feeds

Table 15-2. Suggested Feed Per Tooth for High-Speed Steel Milling Cutters

Material	Face Mills	Helical Mills	Slotting and Side Mills	End Mills	Form Relieved Cutters	Circular Saws
Plastics	0.013	0.010	0.008	0.007	0.004	0.003
Magnesium and Alloys	0.022	0.018	0.013	0.011	0.007	0.005
Aluminum and Alloys	0.022	0.018	0.013	0.011	0.007	0.005
Free Cutting Brasses and Bronzes	0.022	0.018	0.013	0.011	0.007	0.005
Medium Brasses and Bronzes	0.014	0.011	0.008	0.007	0.004	0.003
Hard Brasses and Bronzes	0.009	0.007	0.006	0.005	0.003	0.002
Copper	0.012	0.010	0.007	0.006	0.004	0.003
Cast Iron, Soft (150–180 B.H.)	0.016	0.013	0.009	0.008	0.005	0.004
Cast Iron, Medium (180–220 B.H.)	0.013	0.010	0.007	0.007	0.004	0.003
Cast Iron, Hard (220–300 B.H.)	0.011	0.008	0.006	0.006	0.003	0.003
Malleable Iron	0.012	0.010	0.007	0.006	0.004	0.003
Cast Steel	0.012	0.010	0.007	0.006	0.004	0.003
Low Carbon Steel, Free Machining	0.012	0.010	0.007	0.006	0.004	0.003
Low Carbon Steel	0.010	0.008	0.006	0.005	0.003	0.003
Medium Carbon Steel	0.010	0.008	0.006	0.005	0.003	0.003
Alloy Steel, Annealed (180–220 B.H.)	0.008	0.007	0.005	0.004	0.003	0.002
Alloy Steel, Tough (220–300 B.H.)	0.006	0.005	0.004	0.003	0.002	0.002
Alloy Steel, Hard (300–400 B.H.)	0.004	0.003	0.003	0.002	0.002	0.001
Stainless Steels, Free Machining	0.010	0.008	0.006	0.005	0.003	0.002
Stainless Steels	0.006	0.005	0.004	0.003	0.002	0.002
Monel Metals	0.008	0.007	0.005	0.004	0.003	0.002

Courtesy Cincinnati Milacron Co.

are given in thousandths of an inch per tooth for the various cutters. Multiply the feed per tooth by the number of teeth, and multiply that product by the r/min to determine the feed rate in i/m as follows:

$$i/min = feeds\ per\ tooth \times number\ of\ teeth \times r/min$$

In actual practice, it is better to start the feed rate at a lower figure than that indicated in the table and to work upward gradually until the most efficient removal rates are reached. Too high a feed rate is indicated by excessive cutter wear. A cutter can be spoiled by too fine a feed, as well as by too heavy a feed. A rubbing action, rather than a cutting action, can dull the cutting edge, and excessive heat can be generated. This can be true also in relationship to depth of cut. The first cut on castings and rough forgings should be made well below the surface skin. Milling cuts less than 0.015 inch

Table 15-3. Suggested Feed Per Tooth for Sintered Carbide-Tipped Cutters

Material	Face Mills	Helical Mills	Slotting and Side Mills	End Mills	Form Relieved Cutters	Circular Saws
Plastics	0.015	0.012	0.009	0.00'.	0.005	0.004
Magnesium and Alloys	0.020	0.016	0.012	0.010	0.006	0.005
Aluminum and Alloys	0.020	0.016	0.012	0.010	0.006	0.005
Free Cutting Brasses and Bronzes	0.020	0.016	0.012	0.010	0.006	0.005
Medium Brasses and Bronzes	0.012	0.010	0.007	0.006	0.004	0.003
Hard Brasses and Bronzes	0.010	0.008	0.006	0.005	0.003	0.003
Copper	0.012	0.009	0.007	0.006	0.004	0.003
Cast Iron, Soft (150–180 B.H.)	0.020	0.016	0.012	0.010	0.006	0.005
Cast Iron, Medium (180–220 B.H.)	0.016	0.013	0.010	0.008	0.005	0.004
Cast Iron, Hard (220–300 B.H.)	0.012	0.010	0.007	0.006	0.004	0.003
Malleable Iron	0.014	0.011	0.008	0.007	0.004	0.004
Cast Steel	0.014	0.011	0.008	0.007	0.005	0.004
Low Carbon Steel, Free Machinig	0.016	0.013	0.009	0.008	0.005	0.004
Low Carbon Steel	0.014	0.011	0.008	0.007	0.004	0.004
Medium Carbon Steel	0.014	0.011	0.008	0.007	0.004	0.004
Alloy Steel Annealed (180–220 B.H.)	0.014	0.011	0.008	0.007	0.004	0.004
Alloy Steel, Tough (220–300 B.H.)	0.012	0.010	0.007	0.006	0.004	0.003
Alloy Steel, Hard (300–400 B.H.)	0.010	0.008	0.006	0.005	0.003	0.003
Stainless Steels, Free Machining	0.014	0.011	0.008	0.007	0.004	0.004
Stainless Steels	0.010	0.008	0.006	0.005	0.003	0.003
Monel Metals	0.010	0.008	0.006	0.005	0.003	0.003

Courtesy Cincinnati Milacron Co.

should be avoided. To obtain a good finish, take a roughing cut followed by a finishing cut, with a higher speed and lighter feed for the finishing cut.

In general, feeds can be increased as speeds are reduced. Therefore, the feed can be increased for abrasive, sandy, or scaly material, and for heavy cuts in heavy work. Feed should be increased if cutter wear is excessive or if there is chatter.

Feeds should be decreased for a better finish, when taking deep slotting cuts or when the work cannot be held rigidly. If the cutter begins to produce long, continuous chips, the feed should be decreased.

A good commercial finish can be obtained by using a feed rate that removes a chip of 0.030 inch to 0.050 inch per revolution of the cutter. Finer finishes, of course, necessitate finer feeds, but a feed of 0.015 inch to 0.020 inch per revolution usually produces an excellent finish.

Coolants

Various liquids and compressed air are used to carry off some of the heat generated in machining of metals. The kind of coolant depends on the material being machined.

Cast iron machined at usual speed and feed does not overheat the milling cutters. A liquid coolant should not be used with cast iron because the chips mix with the liquid, resulting in a sticky mass that clogs the cutter teeth and is difficult to remove.

Compressed air can keep the cutter cool and free from chips. The chief disadvantage of compressed air is that too much pressure can scatter the chips and dust, which makes the machine untidy and causes trouble.

Brass can be milled dry. Steel generates considerable heat in machining; therefore, an abundance of coolant should be used. Lard oil makes an excellent coolant. The use of a coolant makes both light and heavy cuts possible at much higher speeds, thereby permitting a smaller cut per tooth, which reduces the stresses on the workpiece and the arbor.

The large volume of lubricant carries away most of the chips, which aids in reducing cleaning on some jigs and fixtures. A lubricant is usually applied to the cutter through piping that has a swivel joint, so that the liquid can be directed to the cutter.

Taking the Milling Cut

A quick check should be taken to make sure that everything is in order before you actually begin to cut metal. Make certain that the work and fixture will clear all parts of the machine and that the cutter will not strike any part of the fixture or holding device. The machine should be well oiled, and there should be an adequate supply of cutting fluid in the reservoir. All table movements that will not be used in the operation should be locked, and all those movements that will be used should be unlocked. The table dogs should be set to throw off the feed after the required length of cut. The dogs should be set to stop the rapid traverse movement before the work reaches the cutter, if rapid traverse is to be used. The starting lever and table control levers should all be in neutral. The spindle speed control should be set to the required speed, and the feed rate should be set at one-half to one-quarter of the feed rate selected for cutting.

After starting the main motor and cutting fluid pump, push over the starting lever and make certain that the cutter is rotating in the correct direction. If it is rotating in the wrong direction, stop the spindle, engage the reverse lever, and restart the spindle. The work should be positioned for a light trial cut before going to full depth. Push the table control lever in the direction the table is to move, and take the trial cut for about ½ inch in length. Then stop the machine, reverse the table, and examine the cut. If the trial cut is satisfactory, increase the feed rate to the full rate selected; take a full-depth cut over the entire workpiece surface, making certain that plenty of cutting fluid is pouring over the point where the cutter engages the metal. Unless there is a very good reason, do not stop the cutter until it has passed entirely over the work; or a depression or scallop will be left on the surface, which is particularly important on finishing cuts.

The work should not be reversed under the cutter without first backing it away slightly. To do so can cause chatter and can cause a jam that can break the cutter and spoil the work. All feed screws have backlash, so a slight backward movement on the handle will not move the table the same amount. Note the starting point, then give the handle at least a half-turn. When the handle is returned to the starting point, the work will be returned to exactly the same place; then the handle can be advanced the required amount for the next cut.

ATTACHMENTS AND ACCESSORIES

A wide variety of standard attachments and accessories can increase the overall usefulness of the standard milling machine. One of the most useful of these is the *universal dividing head* (Fig. 15-12). This device provides a means of holding the work in a chuck, between the centers, or in a collet, and revolving it (indexing) through a desired number of equal divisions. The unit is mounted on the machine table, and the head can be tilted to any angle. The work can be rotated through a full 360° by a worm or bevel gears.

A series of index plates is available for dividing the circle into any desired number of parts. For example, to divide into seven parts, select a circle of holes in the index plate that is divisible by 7, such as 28. As 40 revolutions of the crank are necessary for one

Courtesy Cincinnati Milacron Co.

Fig. 15-12. The universal dividing head permits indexing through any desired number of divisions.

revolution of the work on the spindle, divide 40 by the number of divisions (40 ÷ 7) to give 5⁵⁄₇. Therefore, to obtain seven equal divisions, it is necessary to turn the crank five complete turns and a ⁵⁄₇ turn for each division. As the circle has 28 holes, use ⁵⁄₇ of 28, or 20 holes. So, after each cut, turn the crank 5 complete turns, plus 20 spaces on the 28-hole circle.

A *micrometer table attachment* for jig and die work is shown in Fig. 15-13. This attachment enables the operator to perform boring and recessing operations.

Vertical milling attachments are designed so that vertical milling can be performed on a horizontal machine (Fig. 15-14). These attachments can be used for drilling, boring, cutting T-slots, and other such work.

323

Courtesy Cincinnati Milacron Co.

Fig. 15-13. Micrometer table attachment for use on a milling machine.

SUMMARY

The operator of a milling machine should be aware of certain safety precautions. If proper precautions are taken, danger can be avoided and safe profitable operations can be assured. Never move an operating lever without knowing in advance the action that is going to take place. Stay clear of the milling cutter—do not touch a cutter unless there is a good reason to do so. These are just a

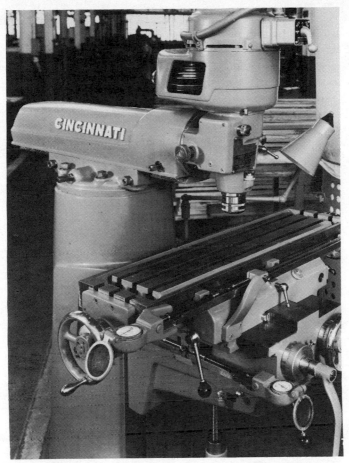

Courtesy Cincinnati Milacron Co.

Fig. 15-14. Milling a dovetail with the vertical milling attachment and swivel vise.

few precautions that should be observed. Keeping the machine clean and oiled cannot be emphasized too strongly. A clean and oiled machine is generally in better mechanical condition than one on which dirt has been allowed to accumulate.

Workpieces must be mounted properly and securely. Clamping devices, such as T-bolts, toe dogs, and screw pins are similar to those used to hold planer workpieces. Always keep cutters clean

325

and sharp. Dull cutters wear more rapidly than sharp ones, and do not last as long as cutter kept in proper conditions.

Proper speed and feed rates are governed by several factors. These factors are material, cutter width, depth of cut, surface finish, power and speed available, and cutting fluid. The amount of material to be removed per minute and the relationship between depth of cut and feed influence cutter speed. The feed rate is the rate at which the work advances past the cutter, which is commonly referred to as inches per minute.

REVIEW QUESTIONS

1. What should the operator of a milling machine do to make the machine a safe piece of equipment?
2. Why should the machine be kept clean and properly oiled?
3. What is means by clearance angle?
4. List some of the mounting methods and devices used with a milling machine.
5. Why is selecting the proper milling machine cutter so important?
6. Why is sharpening a milling machine cutter so important?
7. What is the purpose of the overarm on a milling machine?
8. What is an arbor?
9. Why is the direction of cutter rotation in relationship to the direction of feed so important?
10. What is climb milling?

CHAPTER 16

The Milling Machine Dividing Head

The dividing head (also called index head) is a device that can be used to rotate a piece of work through given angles—usually equal divisions of a circle. A dividing head, in combination with the longitudinal feeding movement of the table, is used to impart a rotary motion to a workpiece for helical milling action, such as in milling the helical flutes of cutters.

Dividing heads are used in milling operations whenever it is necessary to divide a circle into two or more parts (Fig. 16-1). This includes spacings for gear teeth or milling cutter teeth, boring holes in jigs, and other kinds of work that require precision spacing. When a tailstock is used with the dividing head to support the workpiece between centers, the combination setup is called an index center. When it is in use, the dividing head is bolted to the table of the milling machine and becomes part of the equipment of the machine.

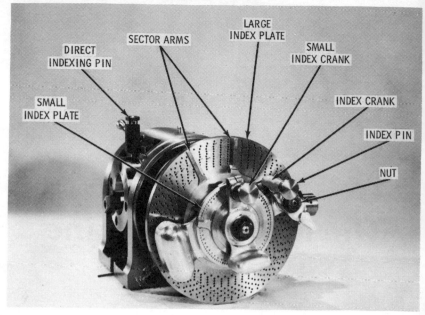

DIRECT
INDEXING PIN

SECTOR ARMS

LARGE
INDEX PLATE

SMALL
INDEX CRANK

SMALL
INDEX PLATE

INDEX CRANK

INDEX PIN

NUT

Courtesy Cincinnati Milacron Co.

Fig. 16-1. Universal wide-range divider for milling machine.

CLASSIFICATION

The types of dividing heads used on milling machines are: (1) plain, (2) universal, and (3) helical. Dividing heads can also be classified as to size; for example, the Cincinnati dividing head is available in three different sizes.

Plain Dividing Head

The basic parts of a plain dividing head are shown in the diagrams (Fig. 16-2 and Fig. 16-3). In a plain dividing head, the spindle rotates about a horizontal axis.

The principal parts of the dividing head are: spindle, worm wheel, worm, index plate, index pin, and sector.

The *worm wheel* is keyed to the *spindle* so that the spindle turns with the wheel. The worm wheel has 40 teeth on most dividing heads.

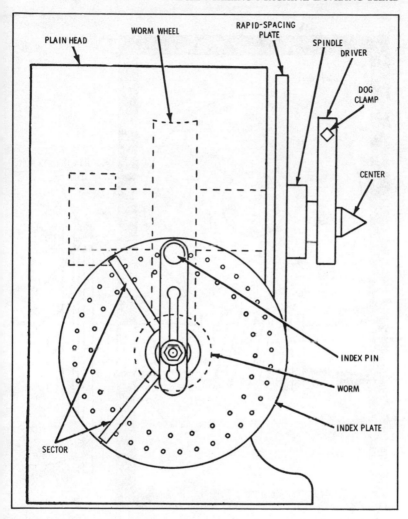

Fig. 16-2. Diagram showing basic parts of a plain dividing head (side view).

The *worm* meshes with the worm wheel. The ratio of the two gears is 40:1; therefore, 40 revolutions of the worm are required to turn the worm wheel one complete revolution.

The *index plate* is one of a set that is provided with the dividing head. The index plate can be removed easily and another plate

329

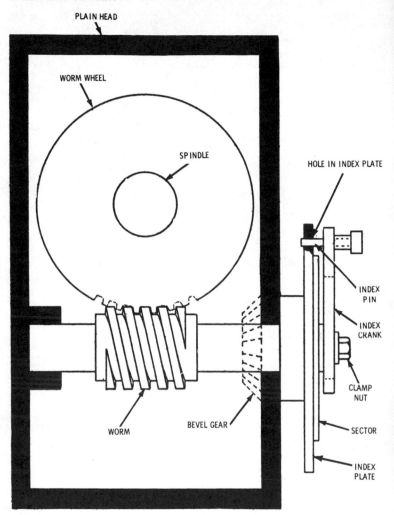

Fig. 16-3. Diagram showing interior parts of a plain dividing head (cross-sectional view.)

substituted as necessary for the desired spacing. Each index plate consists of several circular rows of holes—each circular row having a different number of holes.

The *index pin* is located on the end of the crank, which is attached to the worm shaft. The crank is used to rotate the spindle through the worm gearing. The arm length of the crank is adjustable so that the index pin can drop into any hole in any circular row of holes.

The *sector* consists of two radial arms constructed in such a manner that the angle between them can be changed and locked by a clamp screw in any included angular position. The sector can be used to save time and reduces the possibility of an error in counting the number of holes for each movement of the index pin. In actual operation, the index crank should be adjusted for the correct circular row of holes and the index pin dropped into the correct hole in the row; rotate the sector arm *A* (see Fig. 16-2) against the left-hand side of the index pin, then move the sector arm *B* in the same direction that the index pin is to turn until the correct number of holes is counted between the pin and the sector arm *B*, and lock both sector arms in position. *Never* count the hole in which the index pin has been inserted—this hole is the "zero" hole.

Universal Dividing Head

The universal dividing head differs from the plain dividing head in that the spindle can be tilted, that is, swiveled to any angular position in a vertical plane, within the angular range provided (Fig. 16-4). As shown in Fig. 16-4, the casting that carries the spindle is mounted in a circular guide, forming part of the swivel or universal head. The spindle can be tilted to any angular position *AB* by rotating the spindle casting, which can be clamped in any angular position by tightening the angular clamp bolt. The circular guide is graduated in degrees and fractional degrees.

The universal dividing head is built for a greater range of work than the plain dividing head. For some kinds of tapered work, it is necessary to tilt the spindle at an angle to the table; the universal dividing head is designed for this kind of work. Angular graduations at the top of the housing serve as a guide to setting the spindle at any angle with reference to the horizontal. Fig. 16-5 is an illustration of a milling operation in which the spindle is tilted at an angle.

Helical Dividing Head

The helical dividing head differs from the plain and universal types in that the spindle of the head can be connected to the table

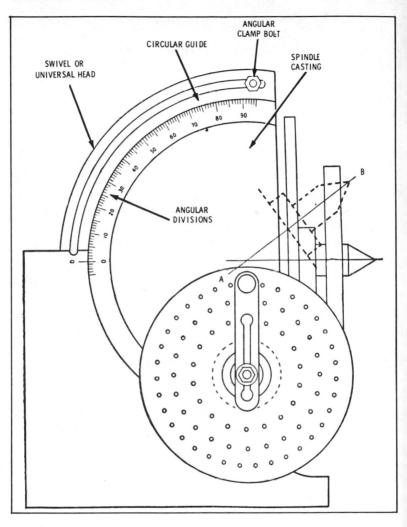

Fig. 16-4. Diagram of the universal dividing head, showing the arrangement for swiveling the spindle (side view).

lead screw by gears so that the work can be rotated as it is moved longitudinally by the table; the two movements are in a definite ratio that is determined by the combination of gears used. These gears are similar to the feed gears on a lathe.

The combination of the rotary and longitudinal movements causes the tool to cut a helix *AB*, as illustrated in the diagram (Fig.

Courtesy Cincinnati Milacron Co.

Fig. 16-5. Setup for the dividing head for milling cams.

16-6). As shown in the diagram, the cutting tool begins at A, cutting the helix from A to B as the tool rotates in the direction indicated by C, while the table is moving the work in the direction indicated by D. The helix cut is either a right-hand or left-hand helix, depending on the direction in which the work is rotated.

The pitch of the helix depends on the rate of rotation of the work with respect to the movement of the table. Milling the flutes of a cast iron rotor is illustrated in Fig. 16-7.

The basic parts of a helical dividing head are shown in Fig. 16-8. A plain dividing head is driven by a gear train from the table feed screw, which converts it into a helical dividing head. A bevel gear (shown in dotted lines in Fig. 16-3 on the shaft to which the worm is attached) is usually free to rotate on the worm shaft. The index plate is fastened to the hub of the bevel gear and is kept from rotating on the shaft by a stop pin in the housing. The index crank is attached to the worm shaft (see Fig. 16-3). If the stop pin is withdrawn and the index pin is inserted in one of the holes in the

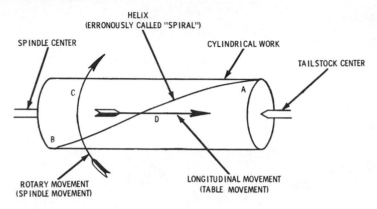

Fig. 16-6. Diagram showing the production of a helix by a combination of rotary and longitudinal feeds.

index plate, the worm shaft and the bevel gear are locked together. Hence, any motion transmitted to the bevel gear is transmitted to the spindle through the gear train, index plate, index pin, index crank, worm, and worm wheel (see Fig. 16-3).

The bevel gear (see Fig. 16-3) meshes with another bevel-gear shaft (Fig. 16-8). Spur gears of various size can be attached to the outer end of the bevel-gear shaft.

The bevel-gear shaft, which drives the spindle, is driven by the table feed screw through the gear train, which consists of the feed screw gear, two stud gears, and the worm gear (see Fig. 16-8). The worm gear is actually a spur gear, but it is usually called the worm gear because it drives the worm in the driving head.

Various changes in the relative movements of the table feed screw and the spindle can be obtained by using gears of different sizes. The diagram shows only a single gear train. Sometimes an idler gear is interposed, as on an engine lathe when the selected gear diameters do not permit direct connection of the feed screw and the bevel-gear shaft.

DIVIDING HEAD AND DRIVING MECHANISM

The Cincinnati Milacron Co. manufactures dividing heads in 10-inch, 12-inch, and 14-inch sizes. these dividing heads are used extensively in milling spiral and helical gears, cams, etc. Motion is

Courtesy Cincinnati Milacron Co.

Fig. 16-7. Milling the flutes of a cast-iron rotor.

transmitted to the spindle in the same manner in all the different sizes of heads. Two types of driving mechanisms can be used on all heads: (1) *enclosed standard lead driving mechanism,* which is standard equipment on universal knee-and-column machines and is available for plain and vertical machines; and (2) *short and long lead attachment,* which is extra equipment.

The dividing head is designed so that its spindle can be swiveled vertically. The spindle is housed in a swivel block; it can be

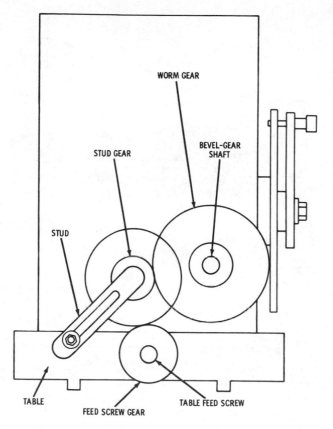

Fig. 16-8. Diagram of a plain dividing head, showing the gear train used in cutting a helix.

swiveled to any angle from five degrees below the horizontal to five degrees beyond the vertical. Thus, bevel gears of any pitch angle can be milled, as well as many other kinds of work that require concentrically spaced slots or holes at an angle to the center line of the workpiece.

Because of the splined shaft drive, the headstock of the dividing head is not required to be placed flush with the end of the table when set up with the driving mechanism. This kind of setup is required for milling scrolls and certain kinds of cams.

336

Setup of Dividing Head and Driving Mechanism

The instructions outlined for setting up the dividing head and driving mechanism should be followed in order:

1. The table of the milling machine and the bottoms of the dividing head and tailstock should be cleaned.
2. Clamp the headstock of the dividing head in the center slot of the table, in a location suitable for the work.
3. Check the spindle of the dividing head with a test bar and indicator to be certain that it is parallel to the table.
4. Depending on the length of the work, clamp the tailstock in proper position.
5. Align the tailstock center with the headstock center.
6. Center and align the cutter with the dividing head or headstock center.
7. Lock the saddle in position.
8. Swing the table to the helix angle that is to be milled (universal machine only). If a universal milling attachment is used on a plain milling machine, swing it to the helix angle that is to be milled.
9. Lock the housing in position (universal machines only).
10. Withdraw the index plate stop. The index plate must be free to revolve with the index pin. *Note:* The stop that engages the notches in the rim of the index plate should be engaged only when the dividing head is used without the driving mechanism.
11. Set up the change gears.
12. Set the sector of the index plate for the proper spacing.
13. Oil the dividing head and the change gears thoroughly.

Note: The use of the power rapid traverse for the table is not recommended when the driving mechanism is connected to the dividing head—especially when the head is equipped with a wide-range divider.

How to select the proper change gears—If it is desired to cut a helix having a 15¼-inch lead, consult the "Table of Leads," and find the lead that is nearest 15¼ inches—15.250 inches, in this instance (Fig. 16-9). The change gears for this lead, 51, 18, 21, and

Lead of Spiral in Inches	A	B	C	D	Lead of Spiral in Inches	A	B	C
15.241	45	17	19	33	15.966	60	17	19
15.256	51	18	21	39	15.983	51	27	33
15.273	42	33	36	30	16.000	48	36	..

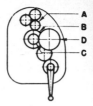

Fig. 16-9. Typical "Table of Leads" for an enclosed driving mechanism.

39 can be used because the lead obtained is near enough to the desired lead for practical purposes.

In some instances the gear C is not listed for some leads. Then, only one intermediate gear is required, and a collar is used to replace gear C (Fig. 16-10).

Setting up the change gears—Remove the bell-shaped cover on the apron at the right-hand end of the table; it is held in position by a large slotted-head screw. Then place the change gears in the positions indicated for the desired lead, making certain that gears B and C are not interchanged. The gears X and Y determine whether the lead is in the right-hand or left-hand direction, as shown in Table 16-1.

After completing the setup, move the table by means of the hand feed crank before engaging the power feed. Remove the hand crank and keep the cover closed while the machine is in operation.

SUMMARY

The dividing head is a device that can be used to rotate a piece of work through given angles, usually equal divisions of a circle. The types of dividing heads used on milling machines are: plain, universal, and helical; dividing heads are generally classified in three different sizes.

In a plain dividing head, the spindle rotates about a horizontal axis. The universal dividing head differs from the plain head in that the spindle can be tilted or swiveled to any angular position in a vertical plane. The helical dividing head is different since the spindle of the head can be connected to the table lead screw by

Fig. 16-10. Enclosed standard lead driving mechanism.

Table 16-1. Determining Direction of Lead

Dividing Head	Right-Hand Helix	Left-Hand Helix
	Remove gear Y Reverse gear X	Gears X and Y, as shown in Fig. 16-10
12-inch and 16-inch Spiral	Gears X and Y, as shown in Fig. 16-10	Remove gear Y Reverse gear X

gears so that the work can be rotated as it is moved longitudinally by the table. The two movements are in a definite ratio that is determined by the combination of gears used.

REVIEW QUESTIONS

1. What is a dividing head?
2. What are the three dividing heads used?
3. What are the basic differences between the dividing heads?

CHAPTER 17

Indexing Operation

As the dividing head (also called index head) is a device used to rotate a workpiece through given angles—usually equal divisions of a circle—the operation in which the spindle of an index head is rotated through a desired angle by means of turning the index crank, which controls the interposed gearing, is called *indexing*.

In most dividing heads, 40 revolutions of the index crank are required for one revolution of a spindle (a ratio of 40:1). Some dividing heads are constructed with a 5:1 ratio.

INDEX PLATES

A great variety of index plates is available for all indexing requirements. Usually, three index plates are provided with the dividing head, each plate consisting of six concentric circular rows of holes; each circular row of holes is designated as an index circle.

The circular row that contains 36 holes is referred to as the "36 circle." A typical set of index plates contains six circular rows of holes, as follows:

> Plate No. 1—15, 16, 17, 18, 19, 20
> Plate No. 2—21, 23, 27, 29, 31, 33
> Plate No. 3—37, 39, 41, 43, 47, 49

The dividing head manufactured by the Cincinnati Milacron Co. is equipped with one standard index plate that has a different series of holes on each side of the plate. The number of holes in each circular row of holes is as follows:

> First side—24, 25, 28, 30, 34, 37, 38, 39, 41, 42, 43
> Second side—46, 47, 49, 51, 53, 54, 57, 58, 59, 62, 66

One of three special index plates can be used to replace the standard plate. Each special index plate has 11 circular rows of holes on each side; each circular row contains the number of holes, as shown in Table 17-1.

Table 17-1. Special Index Plates

Side	Number of Holes in Each Circular Row										
A	30	48	69	91	99	117	129	147	171	177	189
B	36	67	81	97	111	127	141	157	169	183	199
C	34	46	79	93	109	123	139	153	167	181	197
D	32	44	77	89	107	121	137	151	163	179	193
E	26	42	73	87	103	119	133	149	161	175	191
F	28	38	71	83	101	113	131	143	159	173	187

Divisions from 2 to 400,000 can be indexed directly on the "Wide Range Divider" manufactured by the Cincinnati Milacron Co. A large index plate, sector, and crank together with a small index plate, sector, and crank are used on the device. The crank on the small index plate operates through reduction gearing of 100:1 ratio; the ratio between the worm shaft and the spindle is 40:1.

METHODS OF INDEXING

The dividing head can be used for several different methods of indexing. Index tables are provided by the manufacturer of the machine to aid in obtaining angular spacings and divisions.

Direct or Rapid Indexing

Direct indexing is also called rapid indexing, and is used only on work that requires a small number of divisions, such as square and hexagonal nuts. In direct indexing, the spindle is turned through a given angle *without* the interposition of gearing. This is the simplest method of rotating the spindle through a given angle, and is accomplished by turning the spindle by hand.

The index plate is fastened directly to the spindle, so that one complete revolution of the index plate rotates the spindle one complete revolution. Index plates of this type have only a few holes, but they should have as many holes as the number of divisions requires.

For example, and index plate with 24 holes can be used for any number of equal division divisible into 24. The divisions that are possible with a 24-hole index plate are:

$$\frac{24}{24} = 1; \quad \frac{24}{12} = 2; \quad \frac{24}{8} = 3; \quad \frac{24}{6} = 4; \quad \frac{24}{4} = 6; \quad \frac{24}{2} = 12$$

An index plate can be used for any number of divisions that divides the number of holes in the plate equally. A plate with a larger number of holes than is required can be used if the number of holes is an exact multiple of the number of divisions required. In production work, it is safer to use a plate that has exactly the same number of holes as the required number because the possibility of making a mistake in indexing is reduced. A diagram illustrating the basic parts of an index plate that can be used for direct indexing is shown in Fig. 17-1.

As shown in the diagram, the index plate, for the sake of simplicity, consists of 8 index holes. the divisions possible are:

$$\frac{8}{8} = 1; \quad \frac{8}{4} = 2; \quad \frac{8}{2} = 4; \quad \frac{8}{1} = 8$$

Rotating the index plate 8, 4,2, and 1 holes corresponds to angular movements of the plate and spindle of 360, 180, 90, and 45 degrees, respectively.

In all methods of indexing, it should be remembered that the hole from which the latch pin is disengaged when beginning to

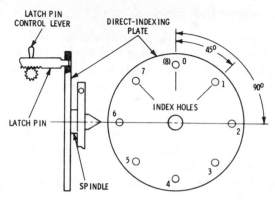

Fig. 17-1. Basic diagram of an index plate, simplified to illustrate the principles of direct indexing.

rotate the index plate is *not counted*—it is the "zero" hole. Actually, the number of spaces between the holes is counted, but it is the usual practice to count the number of holes—except for the beginning hole. The result is the same—the holes are counted, rather than the spaces, because it is much easier for the eyes to follow the holes when counting.

When milling a four-sided workpiece, such as a square nut, the first side is milled with the index pin in the "zero" hole; the second, fourth, and sixth index holes are used, as the remaining sides of the square nut are machined. As noted in the diagram (see Fig. 17-1), the equal spacings—0-2, 2-4, and 4-6 each correspond to a 90° angular movement of the spindle. Similarly, in milling an eight-sided workpiece, the index plate is moved one hole for milling each side.

As mentioned, for simplicity, an index plate with only one circular row of eight holes was used in the diagram (see Fig. 17-1), but index plates are provided that have a greater number of holes. For example, a standard index plate will have three concentric circular rows of holes consisting of 24, 30, and 36 holes, thereby increasing the range of the plate.

Plain Indexing

Plain indexing (sometimes called simple indexing) is an indexing operation in which the spindle is turned through a given angle *with* the interposition of gearing between the index crank and the spindle. This gearing usually consists of a worm on the index crankshaft which meshes with a worm wheel on the spindle. Several turns of the index crank are required for each rotation of the spindle—usually a ratio of 40:1.

The gearing arrangement provides a wide range of divisions or angular movements that are impossible on a direct-indexing plate. As the index crank makes 40 revolutions to each revolution of the spindle, one revolution of the crank provides $\frac{1}{40}$ revolution of the spindle. Like wise, 5, 10, and 20 revolutions turn the spindle one-eighth, one-fourth, and one-half revolutions, respectively.

All dividing or index heads are provided with index plates that have circular rows containing a different number of holes, so almost any fractional turn of the index crank can be accomplished by using the correct index plate.

Use of the dividing-head sector—On the Cincinnati dividing head, a sector, which is concentric with the index plate and crank, is used for plain indexing a fraction of a turn—especially when the procedure is repeated a number of times. The sector enables the operator to locate the index pin in the correct hole without counting the number of spaces that the pin must be moved forward for each division on the work, that is, after the sector has been set for the required number of spaces, as shown in Fig. 17-2. The two

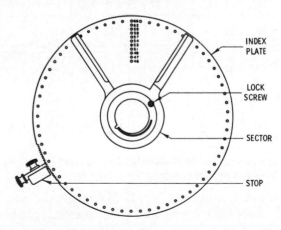

Fig. 17-2. Diagram of dividing-head sector.

345

arms of the sector are set apart at a distance that includes one hole more than the number of spaces that are to be indexed. This factor is sometimes overlooked and can be the source of error in setting up work on the dividing head.

In setting the sector, the narrow edge of the left-hand sector arm should be placed in contact with the index pin. The index crank should be moved the required number of spaces, then dropped into the corresponding hole. Then, the right-hand sector arm should be placed against the pin and the sector arms tightened by means of a lock screw (see Fig. 17-2).

The following example can be used to describe plain indexing. If the correct setting for a 23-tooth gear is desired, first consult the "Index Tables," and note that the 46-hole circle must be used for 23 divisions (Fig. 17-3). Set the index plate so that the side with the 46-hole circle faces the index pin. Set the index pin in any hole in the 46-hole circle, and space the sector for 34 spaces (see Fig. 17-3). Then, for each of the 23 divisions, rotate the index pin through one revolution of the crank, plus the sector spacing, or 34 spaces.

No. of Divisions	Circle	Turns	Spaces	No. of Divisions	Circle	Turns	Spaces	No. of Divisions	Circle	Spaces	No. of Divisions	Circle	Spaces	No. of Divisions	Circle
2	Any	20	....	37	37	1	3	80	24	12	148	37	10	248	62
3	24	13	8	39	39	1	2	82	41	20	150	30	8	250	25
21	42	1	38	56	28	....	20	112	28	10	196	49	10	344	43
22	66	1	54	57	57	....	40	114	57	20	200	30	6	360	54
23	46	1	34	58	58	....	40	115	46	16	204	51	10	368	46
24	24	1	16	59	59	....	40	116	58	20	205	41	8	370	37
25	25	1	15	60	24	....	16	118	59	20	210	42	8	376	47
26	39	1	21	62	62	....	40	120	24	8	212	53	10	380	38
27	54	1	26	64	24	....	15	124	62	20	215	43	8	390	39
28	42	1	18	65	39	....	24	125	25	8	216	54	10	392	49

Fig. 17-3. The correct circular row of holes to use for the desired number of divisions can be obtained from the index table.

The index plate stop (see Fig. 17-2) engages the notches in the index plate, keeping it from rotating. For a spiral or helical milling job, the index plate, sector, and crank rotate as a unit if the dividing head is connected to the driving mechanism. The index plate stop must be disengaged for such a setup. The stop should be engaged only when the dividing head is not connected to the power drive, as when milling spur gears, bolt heads, etc. The stop serves as a safety precaution to prevent errors that would occur if the index plate were moved slightly while indexing.

The stop can also be used to reset work that has been removed for inspection purposes. First, reset the work in an approximate relationship with the cutter. Withdraw the index plate stop, and with the index pin engaged, rotate the crank a sufficient amount to position the work accurately. Re-engage the stop in the notches on the rim of the index plate. Two inches of the circumference of the index plate are notched—the notches have a pitch of 0.060 inch. Thus, moving the index plate one notch is equivalent to rotating the work $\frac{1}{18,460}$ revolution.

How to calculate indexing on the standard driving head—The following rules and example illustrate the procedure for obtaining the maximum number of settings for indexing. As the ratio between the worm and worm wheel on the Cincinnati dividing head is 40:1, then:

1. Divide 40 by the number of divisions required, to give the number of turns or fractional turn of the index crank.
2. If a fraction of a turn is required, the denominator (lower part of the fraction) represents the circle to be used; the numerator (upper part of the fraction) represents the number of spaces in the circle that must be passed over by the index pin.
3. Reduce the fraction to its lowest terms; multiply both the numerator and denominator by the same number until the denominator equals the number of holes in one of the circles on the index plate.

Example: It is desired to calculate all the indexing circles that can be used for three divisions. The following formula can be used:

$$t = \frac{N}{D}$$

in which:
 t = number of completed turns and/or fraction of a turn of the index crank.
 N = number of turns of the index crank for each revolution of

the dividing-head spindle or workpiece. This is equal to 40 turns of the dividing head.

D = number of divisions required in the workpiece.

Thus,

N = 40 and D = 3; therefore, from the above formula:

$$t = \frac{40}{3}$$

$$t = 13\,^1/_3 \text{ revolutions}$$

One-third of a revolution can be obtained by rotating the index pin over one space in a 3-division circle (Rule 2). As a 3-hole index circle is not available, an index circle having a number of holes that can be divided into 3 equal divisions must be used. For example, 8 spaces in the 24-hole circle ($^8/_{24}$ = $^1/_3$), 10 spaces in the 30-hole circle ($^{10}/_{30}$ = $^1/_3$), etc., can be used. The one-third turn can be obtained in each of the following index circles:

1. $^1/_3 \times {}^8/_8 = {}^8/_{24}$ or 8 spaces in the 24-hole circle.
2. $^1/_3 \times {}^{10}/_{10} = {}^{10}/_{30}$ or 10 spaces in the 30-hole circle.
3. $^1/_3 \times {}^{13}/_{13} = {}^{13}/_{39}$ or 13 spaces in the 39-hole circle.
4. $^1/_3 \times {}^{14}/_{14} = {}^{14}/_{42}$ or 14 spaces in the 42-hole circle.
5. $^1/_3 \times {}^{17}/_{17} = {}^{17}/_{51}$ or 17 spaces in the 51-hole circle.
6. $^1/_3 \times {}^{18}/_{18} = {}^{18}/_{54}$ or 18 spaces in the 54-hole circle.
7. $^1/_3 \times {}^{19}/_{19} = {}^{19}/_{57}$ or 19 spaces in the 57-hole circle.
8. $^1/_3 \times {}^{22}/_{22} = {}^{22}/_{66}$ or 22 spaces in the 66-hole circle.

The formula can also be used for indexing *more than 40 divisions*. For example, if it is desired to index 152 divisions:

$$t = \frac{40}{152}$$

As the index plate does not have a circle containing 152 holes, it is necessary to transform the fraction into an equivalent fraction with the denominator equal to the number of holes in one of the circles of the index plate. As the index plate does contain a 38-hole circle, the fraction ($^{40}/_{152}$) can be reduced to $^{10}/_{38}$ by *dividing* both the numerator and denominator by 4 as follows:

$$\frac{40 \div 4}{152 \div 4} = \frac{10}{38}$$

Thus, the 38-hole circle can be used, and the index pin must be moved over a series of 10 holes for each of the 152 divisions into which the work is to be divided.

If it is desired to index *less than 40 divisions* (33 divisions, for example), the fraction then becomes $^{40}\!/_{33}$ ($t = ^{40}\!/_{33}$). The index plate does not contain a 33-hole circle; also, both an 11-hole circle and a 3-hole circle are lacking. As these are the only numbers that can be divided into 33, the transformation to a fraction that can be used must be accomplished by *multiplying* both the numerator and denominator by the same number, rather than by dividing, as in the above example. As the index plate contains a 66-hole circle, both the numerator and denominator can be multiplied by 2 as follows:

$$\frac{40 \times 2}{33 \times 2} = \frac{80}{66}$$

In the equivalent fraction $^{80}\!/_{66}$, the denominator (66) represents the index circle and the numerator (80) represents the number of holes over which the index pin must pass for each division. As the 66-hole circle does not contain 80 holes, the pointer must make one complete turn plus 14 more holes ($80 \div 66 = 1\ ^{14}\!/_{66}$).

Calculation of indexing can be illustrated in a practical milling problem. *Example:* Calculate the indexing for milling a hexagonal nut.

As the gearing is in 40:1 ratio, and applying the formula ($t = N/D$):

$$t = \frac{40}{6} = 6\ ^{4}\!/_{6}\ \text{revolutions}$$

The crank can be rotated exactly $^{4}\!/_{6}$ of a revolution by selecting an index plate with the correct number of holes in one of its index circles. By reducing the fraction ($^{4}\!/_{6}$) to its lowest terms ($^{2}\!/_{3}$), an index circle having a number of holes equally divisible by the denominator (3) can be selected.

Assuming that the index plate has index circles with 37, 39, 41, and 49 holes, the 39-circle can be selected (by inspection) because it is the only circle that has a number of holes that is equally divisible by 3; that is, $39 \div 3 = 13$. Thus, moving the index crank 13 holes is equivalent to:

$$\frac{13}{39} = \frac{1}{3} \text{ revolution of the index crank}$$

As the index crank must be moved ⅔ of a revolution, the crank must be moved (13 × 2) or 26 holes for the fractional part of the 6⅔ revolutions. Therefore, the index crank is turned 6 complete revolutions plus 26 holes for each indexing, when milling the hexagonal nut. Before the milling operation is begun, the crank must be adjusted radially so that the index pin will register with the holes in the 39-hole index circle.

As mentioned, to avoid counting the number of holes for the fractional turn, the sector can be adjusted for the 26 holes. For review, set one arm of the sector to touch the index pin, and set the other arm so that it barely uncovers the 26th hole from the pin—*not* counting the hole engaged by the pin. Then, clamp the sector arms in position.

Compound Indexing

The compound indexing operation is a method of obtaining a desired spindle movement by first turning the index crank to a definite setting as in plain indexing, then turning the index plate itself to locate the crank in the correct position. Normally, the index plate is held stationary by a stop pin, which engages one of the index holes. When the stop pin is disengaged, the index plate can be rotated. Compound indexing can be used to obtain divisions that are beyond the range of the plain indexing system.

Occasionally, none of the index plates provided with the machine will have a number of holes that is suitable for providing the correct number of holes for the required fractional turn of the index crank—this can be overcome by compound indexing. Two kinds of compound indexing can be used, according to the relative movements of the index crank and index plate as: (1) positive compounding, and (2) negative compounding.

Positive compounding—To illustrate compound indexing, assume that the index plate has both a 19-hole and a 20-hole circle. Move the index crank one hole in the 19-hole circle. Disengage the stop pin, and rotate the index plate one hole in the *same* direction in the 20-hole circle—this is positive compounding (Fig. 17-4).

The two movements cause the worm to rotate:

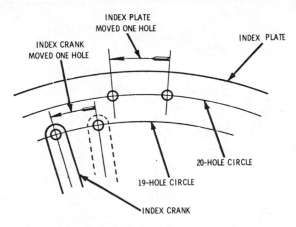

Fig. 17-4. Basic diagram of an index plate, showing positive compounding.

$$\frac{1}{19} + \frac{1}{20} = \frac{20 + 19}{380} = \frac{39}{380} \text{ of a revolution}$$

As the worm-spindle ratio is 40 to 1:

$$\frac{39}{380} \div 40 = \frac{39}{15,200} \text{ of a revolution}$$

Negative compounding—Note that in negative compounding, the movements are in opposite directions (Fig. 17-5). Move the index crank one hole in the 19-hole circle. Disengage the stop pin and rotate the index plate one hole in the *opposite* direction in the 20-hole circle. The resultant rotation of the worm is:

$$\frac{1}{19} - \frac{1}{20} = \frac{1}{380}$$

The worm-spindle ratio is 40 to 1; the spindle movement, therefore, is:

$$\frac{1}{380} \div 40 = \frac{1}{15,200}$$

To index work that requires divisions such as a 69-hole circle, for example, the crank should be turned:

$$\frac{1}{69} \times 40 = \frac{40}{69} \text{ of a revolution}$$

351

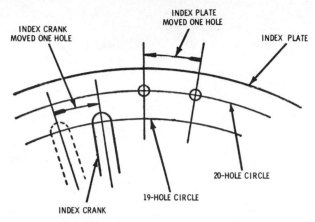

Fig. 17-5. Basic diagram of an index plate, showing negative compounding.

If the index plate contained a 69-hole circle, it would only be necessary to move the crank 40 holes to obtain one spindle revolution. However, in the absence of a 69-hole plate, the same result can be obtained by compounding, using an index plate with both 23- and 33-hole circles.

In compound indexing with these index circles, first move the crank counterclockwise 21 holes in the 23-hole circle; then withdraw the stop pin and move the index plate clockwise 11 holes in the 33-hole circle. The resulting movement of the crank from its original position is:

$$\frac{21}{23} - \frac{11}{33} = \frac{693 - 253}{759} = \frac{440}{759} = \frac{40}{69}$$

As the ratio between the crank and spindle is 40 to 1:

$$\frac{40}{69} \times \frac{1}{40} = \frac{1}{69} \text{ revolution of the spindle}$$

To determine which index circles can be used—The procedure for determining the index circles that can be used can be illustrated for the above example ($\frac{1}{69}$ revolution of the spindle; 40 to 1 ratio), as follows:

1. Resolve into factors the number of divisions required.

$$69 = (23 \times 3)$$

2. For trial and error, choose an index plate. Arbitrarily select two index circles, for example, a 23- and a 33-hole circle.
3. Subtract the number of holes in the smaller index circle from the number of holes in the larger index circle.

$$33 - 23 = 10$$

4. Factor the difference.

$$10 = (2 \times 5)$$

5. Place the two factors (from 1 to 4) *above* a horizontal line.

$$(23 \times 3) (2 \times 5)$$

6. Factor the number of turns of the crank required for one revolution of the spindle.

$$40 = (2 \times 2 \times 2 \times 5)$$

7. Factor the number of holes for each of the trial index circles.

$$23 = (23 \times 1)$$

$$33 = (3 \times 11)$$

8. Place these three sets of factors (from 6 and 7) *below* the horizontal line.

$$\frac{(23 \times 3) \ (2 \times 5)}{(2 \times 2 \times 2 \times 5) \ (23 \times 1) \ (3 \times 11)}$$

9. Cancel the equal factors both above and below the horizontal line.

$$\frac{23 \times 3 \times 2 \times 5}{2 \times 2 \times 2 \times 5 \times 23 \times 1 \times 3 \times 11} = \frac{1}{2 \times 2 \times 1 \times 11}$$

If all factors above the line cancel, the two selected trail index circle can be used. If all factors above the line *do not* cancel, two other index circles must be selected for trial and error, and the procedure must be repeated.

10. To obtain crank movement in a forward direction and plate movement in a reverse direction, multiply the uncancelled factors below the horizontal line.

$$2 \times 2 \times 1 \times 11 = 44$$

11. Thus, the indexing number is 44. This means that to move the spindle $\frac{1}{69}$ revolution, the index crank must be turned forward 44 holes in the 23-hole circle, and the index plate turned 44 holes in the reverse direction, in the 33-hole circle (step 3). The same result can be obtained by making the forward movement in the 33-hole circle and the reverse movement in the 23-hole circle.

The above example is an illustration of negative compounding. Plus (+) and minus (−) symbols are used to indicate the forward and reverse directions of movement in the indexing tables.

In compound indexing, the choice of correct index circles (step 2) usually cannot be solved on the first attempt, as in the example. Normally, two or more selections are necessary before all the factors above the horizontal line can be canceled (step 9).

Compound indexing should not be used when the required divisions can be obtained by simple indexing, because there is a grater possibility for making an error. For this reason, compound indexing has been replaced to a large extent by the differential method. However, a knowledge of compound indexing can be valuable for the machinist.

Sometimes the plain and compound indexing systems can be combined to advantage in an operation such as gear cutting. Every other tooth can be cut by plain indexing, then the spindle positioned to locate the cutter in the center of spaces already cut. Then the remaining spaces are cut by plain indexing as before.

Differential Indexing

In this method of indexing, the spindle is turned through a desired division by manipulating the index crank; the index plate is rotated, in turn, by proper gearing that connects it to the spindle.

As the crank is rotated, the index plate also rotates a definite amount, depending on the gears that are used. The result is a differential action of the index plate, which can be either in the same direction or in the opposite direction in relationship to the direction of crank movement, depending on the gear setup. As motion is a relative matter, the actual motion of the crank at each indexing is either greater or less than its motion relative to the index plate.

In compound indexing, the index plate is rotated manually, with a possibility of error in counting the holes. This is avoided in differential indexing; therefore, chances for error are greatly reduced.

Usually, determination of the gears between the index plate and the spindle can be accomplished by means of an index table that accompanies the machine. The table provides data for both plain and differential gearing; thus, it can be determined quickly whether the latter method can be used.

In the differential indexing operation, the index crank is moved relative to the index plate in the same circular row of holes in a manner that is similar to plain indexing. As the spindle and index plate are connected by interposed gearing, the index plate stop pin on the rear of the plate must be disengaged before the plate can be rotated.

In the gearing hookup, the number of idlers determines whether the plate movement is *positive* (in the direction of crank movement) or *negative* (opposite the direction of crank movement). The gear arrangements are: (1) *simple*, in which the use of one idler provides positive motion, or the use of two idlers provides negative motion to the index plates; and (2) *compound*, in which the use of one idler provides negative movements, or the use of two idlers provides positive movement.

In general, the spindle rotates by means of the worm and worm wheel gearing, as the crank is turned. The index plate is rotated by the gearing between the spindle and the plate; the direction of rotation is either positive or negative, depending on the gear hookup. The total motion or movement of the crank in indexing is equal to its total movement in relationship to the index plate—that is, the sum of its positive motion and its negative motion.

In simple differential indexing (Fig. 17-6), the gear on the worm

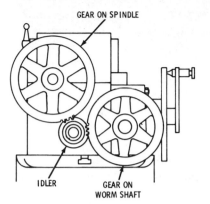

Fig. 17-6. Gearing diagram for
simple differential
indexing.

shaft is the driver, and the gear on the spindle is the driven gear.
The ratio between the number of teeth in each gear determines the
spacing between the divisions. The number of teeth in the idler is
not important, because the idler is used only to connect the other
two gears and to cause the gear on the spindle to rotate in the
desired direction. Two idlers are often used.

In compound differential indexing (Fig. 17-7), the idler is also
used to cause the gear on the spindle to rotate in the desired
direction. Again, the number of teeth in the idler is not important,
but the other gears must have the correct number of teeth for the

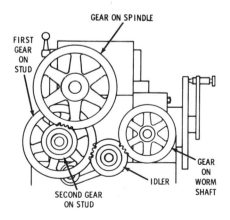

Fig. 17-7. Gearing diagram for compound differential indexing.

desired spacing. The gear on the worm shaft and the second gear on the stud are driver gears, and the first gear on the stud and the gear on the spindle are driven gears.

To select the correct change gears, it is necessary to find the required gear ratio between the spindle and the index plate. Then the correct gears can be determined. The following formulas can be used to determine these gears, in which:

- N is the number of divisions required
- H is the number of holes in the index plate
- n is the number of holes taken at each indexing
- V is the ratio of gearing between the index crank and the spindle
- X is the ratio of the train of gearing between the spindle and the index plate
- S is the gear on the spindle (driven)
- G_1 is the first gear on the stud (driven)
- G_2 is the second gear on the stud (driver)
- W is the gear on the worm shaft (driver)

Then:

If HV is larger than Nn, then:

$$X = \frac{HV - Nn}{H}$$

If HV is less than Nn, then:

$$X = \frac{Nn - HV}{H}$$

For simple gearing,

$$x = \frac{S}{W}$$

For compound gearing,

$$x = \frac{SG_1}{G_2W}$$

Angular Indexing

The operation of rotating the spindle through a definite angle (in degrees) by turning the crank is called angular indexing. Sometimes, instead of specifying the number of divisions or sides required for the work to be milled, a given angle, such as 20° or 45° may be specified for indexing.

The number of turns of the index crank required to rotate the spindle one degree must first be established to provide a basis for rotating the spindle through a given angle. Usually, 40 turns of the index crank are required to rotate the spindle one complete revolution (360°). Thus, one turn of the crank equals 360 ÷ 40 = 9 degrees; or ⅑ turn of the crank rotates the spindle one degree. Accordingly, to index one degree, the crank must be moved as follows:

1. On a 9-hole index plate, 1 hole.
2. On an 18-hole index plate, 2 holes.
3. On a 27-hole index plate, 3 holes.

Example: On an 18-hole index plate, calculate the crank movement needed to index 35°.

As one turn of the crank equals 9°, 35÷ 9 = 3⅞ turns of the crank for 35°. On an 18-hole plate, ⅑ turn equals 18 ÷ 9, or 2 holes, and ⅞ turn equals 2 × 8, or 16 holes. Therefore, to index 35° on an 18-hole plate, 3 turns of the crank plus 16 holes on the plate are required.

It should be noted that:

$$1 \text{ hole in the 18-hole circle} = \tfrac{1}{2}°$$
$$1 \text{ hole in the 26-hole circle} = \tfrac{1}{3}°$$
$$2 \text{ holes in the 18-hole circle} = 1°$$
$$2 \text{ holes in the 27 hole circle} = \tfrac{2}{3}°$$

Example: It is desired to calculate the indexing for 15 minutes (15′). The calculation procedure is as follows:

$$\text{One turn of the crank} = 9°$$
$$9 \times 60' = 540'$$
$$15' = \tfrac{15}{540} \text{ of one turn}$$

Compound Angular Indexing

When the index crank is moved one hole in the 27-hole circle, the spindle rotates 20 minutes (20′); or when the crank is moved one hole in the 18-hole circle, the spindle rotates 30 minutes (30'). The compound method can be used to index angles accurately to 1 minute (1'), as follows:

1. Place a 27-hole plate outside a 20-hole plate so that any two holes register, and fasten them together in position.
2. Turn the plates clockwise three holes in the 20-hole circle; then turn the crank counterclockwise four holes in the 27-hole circle. Thus, the total of these movements is a resultant spindle movement of exactly 1 minute (1') in a clockwise direction.

Block Indexing

This is sometimes called multiple indexing and is adapted to gear cutting. In this operation, the gear teeth are cut in groups separated by spaces; the work is rotated several revolutions by the spindle while the gear teeth are being cut.

For example, when cutting a gear that has 25 teeth, the indexing mechanism is geared to index four teeth at the same time; during the first revolution, six widely separated spaces are cut. During the second revolution, the cutter is placed one tooth behind the previously milled spaces. On the third indexing, the cutter drops behind still another tooth. In this example, the work is revolved four times to complete the gear.

The chief advantage of block indexing is that the heat generated by the cutter (especially when cutting cast iron gears with coarse pitch) is distributed more evenly around the rim of the gear; thus, distortion due to local heating is avoided, and higher speeds and feeds can be used.

SUMMARY

An indexing plate is a plate perforated with variously spaced holes arranged in concentric circles and is used in a milling or similar machine for dividing work, such as spacing out teeth in gear cutting. An index wheel is a circular wheel or disc graduated around its circumference for indicating the angular measurements through which it has been moved. It is used on dividing engines for testing or for adjusting the feed on lathes.

A great variety of index plates is available for all indexing requirements. The dividing head can be used for several different methods of indexing. Index tables are provided by the manufacturer of the machine to aid in obtaining angular spacings and divisions. Direct indexing, also called rapid indexing, is used only on work that requires a small number of divisions, such as square and hexagonal nuts. In direct indexing, the spindle is turned through a given angle without the interposition of gearing.

Plain indexing is an indexing operation in which the spindle is turned through a given angle with the interposition of gearing between the index crank and the spindle. The gearing arrangement provides a wide range of divisions or angular movements that are impossible on a direct-indexing plate.

REVIEW QUESTIONS

1. What is compound, differential, angular, and block indexing?
2. What is an indexing plate and an indexing wheel?
3. What is the purpose of indexing operations?

CHAPTER 18

Shapers

The shaper is a metal-removing machine. The cutting tool is moved in a horizontal plane by a ram having a reciprocating motion, and it cuts only on the forward stroke. The work is usually held in a vise bolted to a box-like table that can be moved either vertically or horizontally.

The size of a shaper is determined by the size of the largest cube that can be machined on it. The shaper was intended originally to produce only plane surfaces—angular, horizontal, or vertical, but it can be used to produce either concave or convex surfaces (Fig. 18-1).

There are two types of shapers—the "crank" and the "geared" shapers. The crank type is the most commonly used, and is made in either standard or universal types. The shaper is used for machining small parts.

BASIC PARTS

The basic diagram of a typical shaper is shown in Fig. 18-2. The diagram illustrates the general assembly of parts and their order.

Base

The base of the machine supports the column or pillar that supports all the working parts. An accurately machined dovetail forms a bearing slide for the ram, and the front of the column is machined to provide slides or ways for the up-and-down movement of the crossrail.

Crossrail

The crossrail is a heavy casting attached to the base by means of gib plates and bolts. The top surface of the crossrail is accurately machined to provide a smooth surface for the saddle as it is fed

Courtesy South Bend Lathe, Inc.

Fig. 18-1. A seven-inch bench shaper.

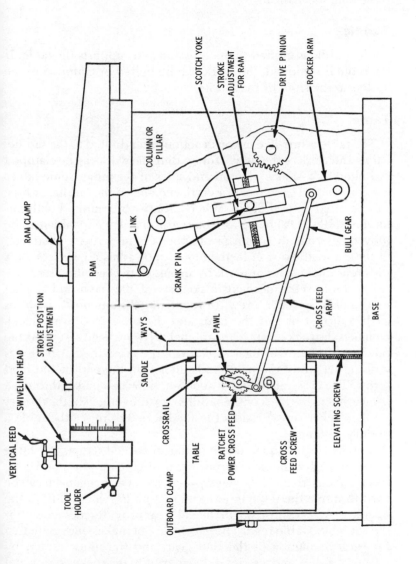

Fig. 18-2. Basic diagram of a typical shaper.

back and forth. The crossrail contains the table elevating and traversing mechanisms.

Saddle

The saddle is gibbed to the crossrail and supports the table. If the table is removed, the work can be bolted or clamped to the T-slots in the front of the saddle.

Table

The table is box-like in construction and is bolted to the saddle. Either the work or the swivel for holding the work can be clamped or bolted to T-slots in the top and sides of the table. Some tables are constructed to swivel in a vertical plane on the saddle, or they can be tilted up and down. When a shaper is equipped with this kind of table, it is known as a universal shaper. The table can be moved horizontally by means of either a rapid traverse hand feed or the automatic power feed, and it can be adjusted vertically for different thicknesses of work by means of the elevating screw.

The table feed mechanism consists of the pawl and ratchet mechanism, a rocker screw for regulating the amount of feed for each return stroke of the ram, and the cross-feed arm, which connects the rocker arm to the ratchet drive and lead screw. As the ratchet screw moves back and forth, the cross-feed arm moves with it, actuating the pawl and ratchet on the end of the feed screw. The crossrail holds the feed screw, which passes through a bronze nut in the back of the saddle. Any movement of the ratchet by the rocker screw causes the feed screw to move the table a definite distance.

A crank fits the squared end of the cross-feed screw. The table can be fed just as fast as the operator desires to turn the crank, or a lever that controls a high-speed power traverse mechanism can be used to move the table in either direction in relationship to the operator. An eccentric slot in the bull gear provides movement for the automatic table feed. The eccentric slot makes one revolution for each revolution of the bull gear, and transmits motion by means of a shoe and a series of lever arms to the rocker screw and cross-feed arm, so that the cross-feed arm moves backward and forward in each revolution of the bull gear.

364

The automatic power feed is engaged by first feeding the work to the cutting tool by hand feed. The pawl is turned to engage the teeth on the ratchet feed screw; as the power feed mechanism is always in motion while the shaper is running, the pawl will be moved back an forth, driving the ratchet in one direction or the other, depending on how the pawl contacts the ratchet. On the return stroke of the ram, the feed screw is moved, causing the table to be fed toward the cutting tool in preparation for the cutting stroke.

Ram

The ram is a strong and rigid casting that is actuated back and forth horizontally in the dovetail slide by means of the rocker arm and the crank pin. The ram contains a stroke positioning mechanism and the downfeed mechanism.

The toolhead slides in a dovetail at the front of the ram by means of T-bolts. It can swivel from 0° to 90° in a vertical plane. The toolhead can be raised or lowered by hand feed for vertical cuts on the workpiece.

Bull Gear

The bull gear is mounted on the column and is driven by a pinion that is moved by the speed-control mechanism. A radial slide, carrying a sliding block into which the crank pin fits, is anchored to the center of the bull gear. The position of the sliding block is controlled by a small lead screw connected to the operator's side of the shaper by means of bevel gears and a squared-end shaft. The location of the sliding block with respect to the center of the bull gear governs the length of stroke of the ram. The farther apart the two centers are located, the greater is the length of stroke.

OPERATING THE SHAPER

Several factors can affect the efficient operation of the shaper. Some of these factors are selection of the cutting tool, holding the work, adjustment of the work, adjustment of the stroke, and selection of proper feed and cutting speed.

Cutting-Tool Selection

The shank of the cutting tool should be in a plane perpendicular to the line of motion of the cutting tool; therefore, the clearance angle should remain constant. The clearance angles on cutting tools for the shaper are not as large as those on cutting tools for the lathe; also, the shaper requires a shorter cutting tool than the lathe, so that the tendency to dig into the work is avoided.

Cutting tools that are commonly used on the shaper are shown in Fig. 18-3. All these tools are available as either straight or bent cutting tools. Roughing, finishing, side roughing, and side finishing tools are most commonly used. Special toolholders with inserted-blade cutting tools can be used effectively on the shaper. Various special cutting tools are also used to perform certain operations. For example, one of these special cutting tools is used for cutting T-slots.

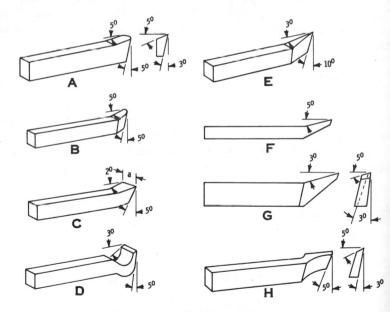

Fig. 18-3. Cutting tools commonly used on the shaper are: (A) right-hand roughing, (B) round nosed, (C) squared nosed, (D) and goose necked. (E) and (F) are views of a right-hand down-cutting side tool; (H) is a left-hand side facing tool.

Holding the Work

As shaper operations are concerned chiefly with small work-pieces, the work is usually held in either a vise or a chuck. These holding devices are auxiliary to the table or machine.

Work bolted directly to the table—It is important to place the stops (T-bolts) properly. If the work were laid on the table without the stops to resist the thrust of the cutting tool on the cutting stroke, the cutting tool would "push" the work rather than cut it.

Large pieces of work that cannot be held in a vise should be fastened to the table. In fastening the work to the table an important consideration is to tighten the clamp bolts no more than is necessary to hold the work firmly on the table to avoid distorting or springing the work. As the cutting tool does not tend to elevate the workpiece, extreme tightening of the bolts is not necessary.

T-bolts having a head of correct size to fit the table slots are commonly used. A complete bolt assembly consists of the bolt, nut, and washer (Fig. 18-4). The bolt should have a thread of sufficient length for work of various thicknesses.

Fastening with clamps—A complete clamping unit consisting of the bolt assembly, clamp, and fulcrum block is shown in Fig. 18-5. In a clamping arrangement, the lever principle should always be considered. Thus, the work should be positioned on the table close

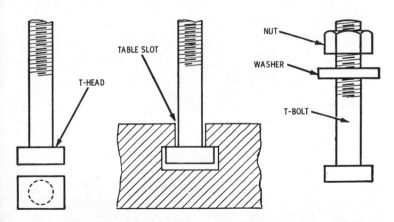

Fig. 18-4. T-head bolt (left) and detail of table showing T-head bolt in the T-slot (center). T-head assembly consisting of bolt, washer, and nut is shown (right).

to the table slot so that the bolt will be as far from the fulcrum block as possible: that is, the bolt should be nearer the work than the block (see Fig. 18-5). Correct and incorrect methods of clamping are shown in Fig. 18-6. It is important that the machinist remember to place the bolt near the work and to tighten the bolts just enough to anchor the work, but not enough to spring the work.

If a number of clamps are used, turn all the bolts until the work is fastened lightly; then tighten the bolts in a staggered sequence to distribute the stresses brought onto the work by clamping (Fig. 18-7). It is also important to select a fulcrum block of the same thickness (or height) as that of the work being clamped to the table. This gives the clamp an even bearing on the work, and the work will be held more securely (Fig. 18-8).

Stop pins—Two types of stop pins are the hole type and the inclined slot type (Fig. 18-9). A stop pin can be anchored in a hole or slot in the table; and the screw can be forced directly against the work, or it can be used to force another device, such as a toe dog, against the work.

Stop pins and toe dogs—A toe dog is a holding device that is similar in shape to a center punch or cold chisel, and it is designed to be forced against the work by a stop pin. Stop pins are used in combination with toe dogs to hold thin work. The pointed and

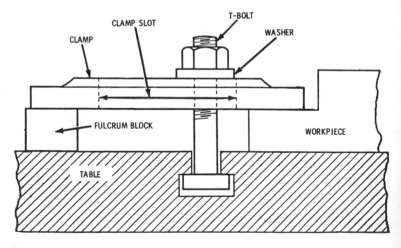

Fig. 18-5. Clamping unit for holding workpiece.

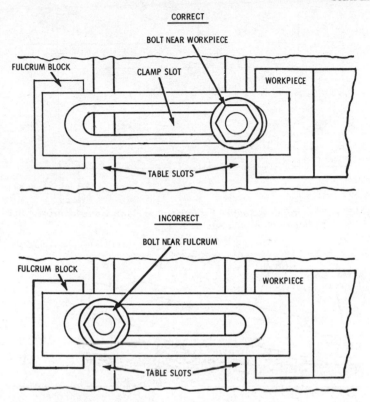

Fig. 18-6. Correct (top) and incorrect (bottom) methods of setting up a clamping unit.

blade types of toe dogs are shown in Fig. 18-10, and their application is shown in Fig. 18-11. Stop pins of either the hole type or the slot type can be inserted in the table on each side of the work, and the dogs can be forced against the work by tightening the stop pin screws. Note that the end of the stop pin projects into the short bore in the end of the toe dog, connecting the two parts (see Figs. 18-10 and 18-11).

The work is pressed against the table because of the annular position of the dogs; however, the angle between the screw and the dog should not be too great. The work should be anchored by an ample number of combined stop pin and toe dog units so that they can take the thrust and secure the work to the table.

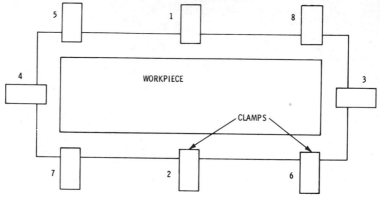

Fig. 18-7. The proper sequence for tightening clamp bolts for proper distribution of clamping stresses.

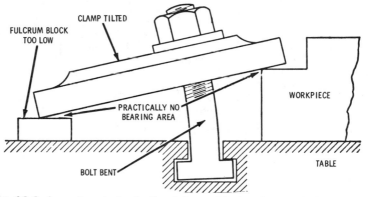

Fig. 18-8. Importance of selecting fulcrum blocks of proper height to hold the work securely.

Stop pins and table strip—A table strip can be used in combination with stop pins to hold work that is thick enough to project above the strip and stop pin screw (Fig. 18-12). The table strip has a tongue on one side that fits in the table slot; this positions the work parallel to the travel of the cutting tool. A method of fastening the work by means of stop pins and table strip is shown in Fig. 18-13. The table strip is held in the slot by two T-bolts. The tongue side of the strip projects downward into the slot. The stop pins are placed in the table holes, and the pins are turned firmly against the work. An adequate number of stops for taking the thrust of the cutting tool should be provided.

370

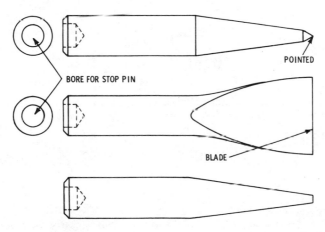

Fig. 18-9. The hole-type (left) and the slot-type (right) stop pins.

Fig. 18-10. The pointed (top) and the blade (center and bottom) types of toe dogs.

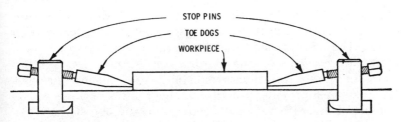

Fig. 18-11. Application of stop pins and toe dogs for holding thin work.

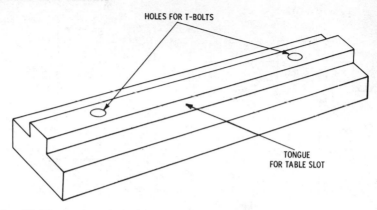

Fig. 18-12. Bottom view of the table strip, showing the tongue and the bolt holes.

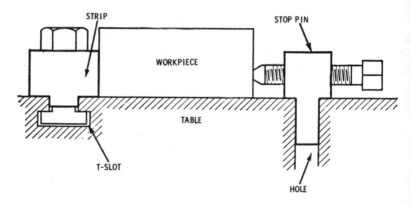

Fig. 18-13. Application of a table strip and stop pins for fastening a workpiece.

Braces for tall castings—Workpieces that project some distance above the table can be machined with the aid of braces, which are used to hold the work and to take the thrust of the cutting tool (Fig. 18-14). The brace is used to aid the clamp (lower stop) in resisting the tendency of the casting to move upward in a circular path. Thus, the work is stop anchored on both the upper and the lower flanges, which provides a rigid fastening for the workpiece. The number of braces required depends on the size and shape of the workpiece; this also applies to the number of clamps and stop pins required.

Holding circular workpieces—If a length of shafting is to be splined or key seated, it must be fastened rigidly to the table, as shown in Fig. 18-15. In this setup, a special table strip with an inclined edge A and a wedge block are used to hold the workpiece. As with an ordinary table strip, the tongue provides parallelism,

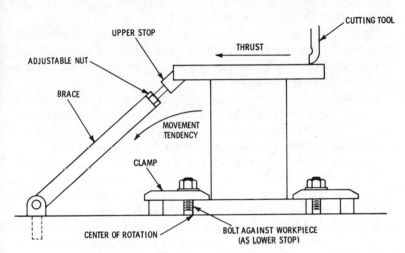

Fig. 18-14. A method of employing a brace to hold a piece of work that projects above the table.

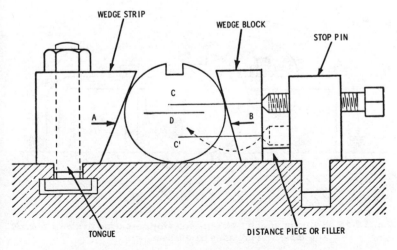

Fig. 18-15. Setup for holding circular or round pieces of work.

373

and a wedge block *B* with a distance piece and a stop pin completes the setup.

The axis *C* of the stop pin should be higher than the center *D* of the workpiece. If these conditions are provided, the workpiece will be clamped firmly to the table. If the axis *F* of the stop pin were placed below the center of the workpiece, the wedge block would rotate (indicated by *E* in Fig. 18-15).

On production work, the special holding devices, such as the strip and block, must be designed to provide the most efficient clamping of the workpiece. A stop should always be included to resist the thrust of the cutting tool.

Another method of clamping a shaft uses the table slot instead of a strip to provide parallelism (Fig. 18-16). This method can be used on tables having transverse slots. Here again, the number and kind of holding devices for a given setup is determined by the size, shape, and any other special characteristics of the workpiece.

Irregularly shaped work bolted indirectly to the table— Numerous fixtures can be used to set up castings having odd shapes for machining. Any device that can be attached or "fixed" to the table for holding and positioning the workpiece is considered to be a fixture. The angle plate is a fixture; one side is bolted (fixed) to the table, and the workpiece is bolted to the other side (Fig. 18-17).

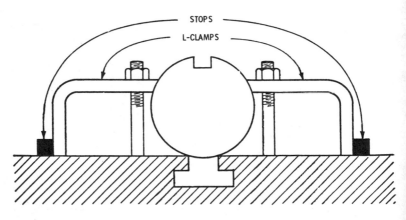

Fig. 18-16. Setup in which L-clamps and stops are used to clamp a shaft. The table slot provides parallelism.

Fixtures are valuable aids in production work as the successive castings not only can be held in position for machining, but each casting can be held in the same position, when bolted to the fixture, as the previous ones. If the casting projects from the angle plate, provision can be made for supporting it on wedge blocks to prevent springing. The wedge blocks should not be tightened enough to spring the casting, but they should be tightened enough to provide support. Springing can be prevented by checking the setup with a surface gage. Note that the angle plate acts as a stop (see Fig. 18-17).

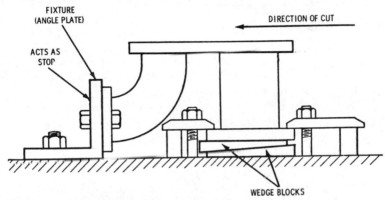

Fig. 18-17. Fixture (angle plate) used to hold an irregularly shaped casting. End support with clamps and wedge blocks is also required to prevent springing of the work.

Work held by a vise—A vise has only two jaws. One jaw is stationary and the other jaw is adjustable. A chuck has more than two jaws, all of which are adjustable. A vise is often erroneously called a chuck. A machine vise is designed for attachment to a machine table, and consists of a base, a table, and both a stationary and an adjustable jaw.

Two general types of vises are the plain (single and double screw) and universal vises. Basic parts of a plain vise are illustrated in Fig. 18-18. The base of a machine vise is constructed with a tongue that fits into the table slot, and it has lugs or open holes for bolting to the shaper table with T-bolts. Also, the base has a circular gage, graduated in degrees, to indicate any angular posi-

375

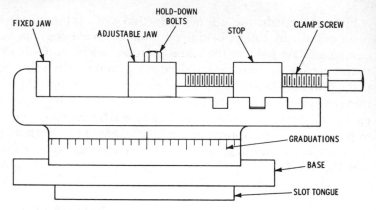

Fig. 18-18. Basic parts of a plain vise.

tion. The base has a stationary jaw at one end and ways for movement of the adjustment jaw; it also carries the clamp screw.

When fastening work in the vise, the movable jaw should be brought near its final position for engagement of the workpiece. The jaw holddown bolts should be tightened tightly; otherwise the movable jaw and the work can tilt (Fig. 18-19).

Workpieces that project above the vise jaws ordinarily cause no problem in holding. A thin piece of work can be placed on parallel strips to project it above the vise jaws so that the jaws will not interfere with the cutting tool (Fig. 18-20). If the workpiece is long and narrow, the vise should be turned so that the cutting tool can be used for longer strokes, rather than shorter strokes.

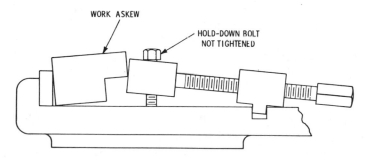

Fig. 18-19. Result of failure to tighten the holddown bolts on the vise.

376

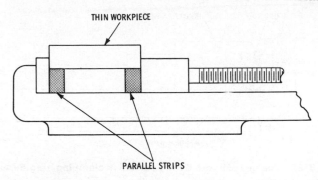

Fig. 18-20. The use of parallels to elevate the workpiece above the vise jaws for machining.

Round work that is too small in diameter to be machined while resting on the ways of the vise can be held in horizontal alignment (parallel with the vise table) by means of V-blocks, as shown in Fig. 18-21. If an irregularly shaped piece of work having a concave side were clamped in the vise, the jaw on the concave side would not be held properly, and the piece would tilt. This can be avoided by using a round rod to aid in holding the work (see Fig. 18-22).

The movable jaw of a universal vise can be swiveled to hold an angular workpiece. A stop can be used to prevent the work from shifting sideways if the side of the piece has too large an angle (Fig. 18-23).

When the movable jaw of a vise is forced against the work, the

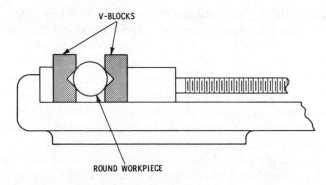

Fig. 18-21. The use of V-blocks to hold round stock in the vise.

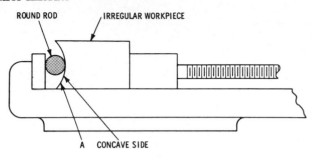

Fig. 18-22. The use of a round rod as an aid in clamping irregularly shaped work.

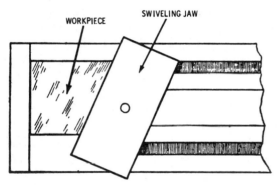

Fig. 18-23. Application of the swiveling jaws of a universal vise to hold an angular workpiece.

jaw tends to lift upward, tilting the work as the jaw is lifted. This can be avoided by using a round rod. If the jaw should lift upward, the workpiece would remain in its original position. (Fig. 18-24).

Special fixtures—Special fixtures for a given workpiece can be designed to hold irregularly shaped castings. The fixture must position and support the casting in the correct places to prevent stresses causes by uneven clamping, which tends to spring the work. The special fixtures should be constructed so that they can be attached to the work with a minimum loss of time.

Workpiece mounted between centers—An attachment similar to the lathe headstock and tailstock can be used to mount some kinds of work that cannot be completely machined on a lathe. The headstock spindle does not rotate, except for turning to any

378

desired angular position as determined by an index head (Fig. 18-25).

To set up the work, the same centers that were used on the lathe should be used on the shaper. A tongue on the bottom of the attachment fits the table slots of the shaper; this provides parallelism and alignment. However, the alignment should be checked before the machine operations are begun.

In the setup (see Fig. 18-25) the headstock spindle can be rotated by turning the handle of the worm gear. The index plate can be

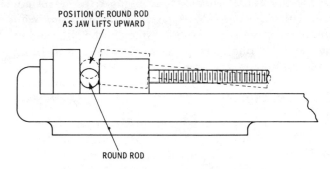

Fig. 18-24. Application of a round rod to prevent tilting of the workpiece should the vise jaw lift upward.

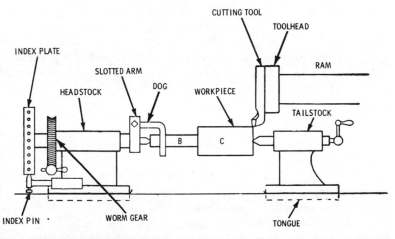

Fig. 18-25. Basic diagram of a setup for mounting workpieces between shaper centers.

locked in any position by engaging the index pin with the corresponding hole in the circumference of the index plate.

For example, a set of five index plates can have 44, 52, 56, 90, and 96 holes. The index plate that is used must have a number of holes that is evenly divisible by the required number of angular positions that are necessary to machine the piece. For instance, the section B (see Fig. 18-25) has been turned in the lathe, and the section C requires six sides that are to be machined in the shaper.

After selecting the index plate with 90 holes (90 + 6 = 15), set the index plate so that the index pin engages the "zero" hole in the plate, and machine one side of the work to the finished dimension. The angular setting for the second side can be obtained by disengaging the index pin, turning the worm gear handle to rotate the index plate 15 holes, and engaging the index pin in the hole.

Adjusting the Work

After the workpiece has been mounted and aligned on the shaper table, the ram should be moved outward until the tool post is over the work. The tool should be placed in the tool post with as little overhang as possible. The table should be elevated until the workpiece clears the ram a distance of about one inch. The tool slide should project only a short distance below the bottom of the shaper head; that is, the overhang of the slide should be not more than 1½ inches. If more overhang is necessary, the cutting tool, rather than the slide, should be set for the extra overhang.

The cutting tool should be clamped firmly, making certain that it clears the work. Then the machine can be started to note the position and length of stroke.

Stroke Adjustment

The ram should be adjusted to provide the proper length of stroke and to provide the proper position of travel over the workpiece. The stroke adjustment screw (see Fig. 18-2) can be turned to produce either a shorter or a longer stroke. A handle is provided on most machines for the square head of the adjusting screw. Unless the operator is familiar with the machine, trial and error can determine the direction of turning the stroke adjustment screw for either a longer or a shorter stroke. The stroke should be long enough for the cutting tool to clear the work by not less than ¼ inch on the forward or cutting stroke; and it should clear by ½ inch behind the workpiece on the return stroke, so that the automatic feed can function before the cutting tool contacts the work for the forward stroke (Fig. 18-26). The automatic feed should function at the end of the return stroke, rather than at the end of the forward or cutting stroke.

If, after proper length of stroke has been obtained, the travel of the cutting tool is not positioned correctly over the work, the position of the ram should be changed. The ram clamp should be loosened, and the ram positioning adjustment turned until the cutting tool clears the work by a distance of ¼ inch beyond the workpiece on the forward stroke and ½ inch behind the workpiece on the return stroke (see Fig. 18-2). Then the ram clamp should be tightened. The operator should remember to adjust the length of stroke before adjusting the position of travel over the workpiece.

Head Adjustment

Movement of the cutting tool and the tool slide is controlled by the vertical feed handle at the top of the head. The handle is attached to the down-feed screw inside the slide. A micrometer dial graduated in thousandths of an inch can be used to make accurate adjustments of the slide and cutting tool. Back lash or lost motion between the threads of the down-feed screw and the nut must be removed before the dial can be set accurately.

The toolhead should be positioned at an angle in shaping a dovetail, and in certain other setups. The operator should be careful not to run the ram back into the column while the slide is set

Fig. 18-26. Front and back positions of the ram in relationship to the workpiece.

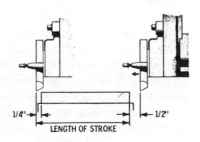

1/4"   LENGTH OF STROKE   1/2"

at an extreme angle, as the slide will strike the column when the ram moves back on the return stroke.

The clapper box should be set so that the top is slanted slightly away from the cutting edge of the tool (Fig. 18-27). Then the tool can swing away from the work on the return stroke of the ram, thus protecting the cutting edge of the tool.

Taking the Cuts

The operator should decide on the number of cuts necessary, depending on the amount of metal to be removed. Usually, one or two roughing cuts and one or two finishing cuts are required, depending on the nature of the work and the finish desired.

The ram should be positioned over the work and the cutting tool adjusted by the vertical feed handle so that it almost touches the work. By means of the hand cross-feed screw, move the table to the right or left (according to the cutting tool in use) until the workpiece is in position for beginning the cut—the tool should be far enough from the work that one or two strokes can be made before the tool begins to cut metal.

Then set the cutting tool to the proper depth for the first cut by means of the vertical feed handle. The first cut should be deep enough to barely remove the scale, thus truing the surface and enabling the operator to check the setting of the cutting too. Then the necessary roughing and finishing cuts can be taken on the workpiece.

Selecting the Proper Feed

In shaping, the feed is the distance the work is moved toward the cutting tool for each forward or cutting stroke of the ram. Either the hand feed or the automatic feeding mechanism can be used. In setting the automatic feed, the amount of feed will depend on the kind of metal being cut. The time required to complete the work with a given finish is determined by the amount of feed. Both the kind of metal and the type of job must be considered in selecting the feed.

If a softer metal is used, a heavier feed can be used (0.030 to 0.060 inch). A finer feed must be used on harder and tougher

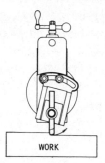

Fig. 18-27. Turn the top of the clapper box away from the cutting edge of the tool to permit the tool to swing away from the work on the return stroke.

metals. (0.005 to 0.015 inch). A feed of 0.062 ($\frac{1}{16}$) inch can be used on cast iron.

Of course, the capacity of the shaper must be considered. Heavier feeds can be used on a large heavy-duty machine than on a small and lightly constructed shaper.

Selecting the Proper Cutting Speed

The speed of the shaper is the number of cutting strokes made by the ram during one minute of operation. This is governed by the speed of the bull gear. The number of strokes made by the ram in one minute remains constant for a given speed of the bull gear whether the stroke is long or short. The cutting speed is changed by changing the rate and amount of rocker-arm movement.

The speed of the cutting tool is the average rate of speed attained when the shaper has been adjusted to make a given number of cutting and return strokes of a given length in one minute. The rate of speed, then, is determined by the time or fractional part of a minute required for the cutting stroke and the distance or total length in feet of the cutting stroke.

Cutting speed is determined by the total distance that the tool travels during the cutting strokes made in one minute, and the ratio of cutting-stroke time to return-stroke time. In most shapers, about $1\frac{1}{2}$ times as much time is required for the cutting stroke as for the return stroke, giving a ratio of 3:2. Thus, the cutting stroke requires $\frac{3}{5}$ of the time for the cycle (a complete cycle consists of one cutting stroke and one return stroke).

If the length of stroke in inches and the number of strokes per minute are given, their product will give the number of inches of metal cut during one minute of operation. As cutting speed is expressed in feet per minute, it must be multiplied by 12.

Divide the distance (in feet) by ⅗, as the tool cuts during ⅗ of the cycle (distance divided by time equals rate). To shorten the calculation use the formula:

$$\text{Cutting speed} = 0.14. \times N \times L$$

in which,

N = strokes per minute
L = length of stroke

Example: Find the cutting speed of the tool when the shaper makes 60 strokes per minute, and the cutting stroke is 10 inches long.

$$\text{Cutting speed} = 0.14 \times N \times L$$
$$= 0.14 \times 60 \times 10$$
$$= 84 \text{ feet per minute}$$

To determine the number of strokes per minute at which the shaper should be run, with the cutting speed of a metal given in a table, use the formula:

$$\text{Strokes per minute} = \frac{\text{cutting speed}}{0.14 \times \text{length of stroke}}$$

Example: Find the number of strokes per minute at which the shaper should be run to machine a piece of copper having a cutting speed of 80 ft. per minute (from table), and the stroke adjusted to 6 inches.

$$\text{Strokes per minute} = \frac{\text{cutting speed}}{0.14 \times \text{length of stroke}}$$
$$= \frac{80}{0.14 \times 6\,\text{inches}} = \frac{80}{0.84}$$
$$= 95.238 \text{ strokes per minute}$$

In most instances, the shaper is operated at a speed that is too slow. Therefore, the beginner should determine the speed at which the shaper should be operated and work as near the speed as possible. Cutting speeds for shaping the common materials are given in Table 18-1. The length of stroke can be obtained from the machine.

Table 18-1. Shaper Cutting Speeds

Materials	CUTTING SPEED (ft. per minute)		LUBRICANT	
	H.S.Tools	C.S.Tools	Roughing	Finishing
Aluminum	125	50	Dry	Kerosene
Brass	Highest possible	35	Dry	Dry
Cast iron (hard)	35	20	Dry	Soda water
Cast iron (soft & medium)	60	30	Dry	Dry
Copper	80	40	Dry	Dry
Steel (hard)	35	15	Cutting compound	Lard oil
Steel (soft)	Highest possible	35	Cutting compound	Soda water

SUMMARY

The shaper is a metal-removing machine with cutting tools that move in a horizontal plane in a reciprocating motion. The size of a shaper is generally determined by the size of the largest cube that can be machined on it.

There are two types of shapers—the crank and the geared type. The crank type is the most commonly used and is made in either a standard or universal type. The shaper was intended originally to produce only plane surfaces, such as angular, horizontal, or vertical, but it can be used to produce either concave or convex surfaces as well.

Basic parts of the shaper are base, crossrail, saddle, table, ram, and bull gear. The base naturally supports the machine and all working parts. The crossrail contains the table elevating and traversing mechanism and is attached to the base by means of gib plates and bolts.

Many factors can affect the efficiency of operation of the shaper. Some of these factors are selection of the cutting tool, holding the work, adjustment of the work, adjustment of the stroke, and the selection of the proper feed and cutting speed.

REVIEW QUESTIONS

1. Name the two types of shapers.
2. Of the two types of shapers, which is most commonly used?
3. What is the basic operation of the shaper?

CHAPTER 19

The Planer

The planer is a machine tool used to machine flat or plane surfaces on work that is fastened to a reciprocating table. The surfaces can be horizontal, vertical, or angular; the planer can also be used to form irregular or curved surfaces.

The planer differs from the shaper in that the worktable moves back and forth with a reciprocating motion while the cutting tool is held stationary, except for transverse movement. The planer is used for the same purpose as the shaper; but it can handle much larger work, and heavier cuts can be taken. The planer is used on work that is too large or otherwise impossible to machine on a shaper. Some typical examples of planer work are: machine bases, machine covers, and column supports.

TYPES OF PLANERS

One of the more common types of planers is the *double-column planer*. Two columns support the crossrail and house the elevating screws and controls for the machine.

The *open-side planer* has only one column or housing to support the crossrail and toolheads. One advantage of this type of planer is that workpieces of irregular shape can be handled with the workpiece extending outward over the side of the table. The working parts are essentially the same as on the double-column planer.

The rated size or capacity of a planer as given by the manufacturer refers to the width, height, and length of the largest workpiece that can be machined. For example, a planer (24 × 24 × 48) indicates that the planer can machine a piece of work 24 inches wide, 24 inches high, and 48 inches in length.

BASIC PARTS

The most important parts of a planer are the bed, columns, crossrail, table, toolhead, and table drive (Fig. 19-1 and 19-2). The working parts are essentially the same on both double-column and open-side planers.

Bed

The bed is an extremely heavy, rigid casting. The table slides in accurately finished ways on the bed. The bed also supports the columns and all moving parts of the machine.

Columns

On a double-column planer, two columns rise vertically at the sides of the machine. They support the crossrail and house the elevating screws and controls for the machine. The open-side planer has only one column.

Crossrail

The crossrail carries the saddle and toolhead. It is a rigid casting and provides guides for transverse travel of the saddle. The crossrail also contains the feed rod and screw for controlling the movement of the cutting tool.

Vertical screws located in each of the columns provide a means of support and a means of adjusting the crossrail vertically. It is of utmost importance that the crossrail, when clamped, be parallel to

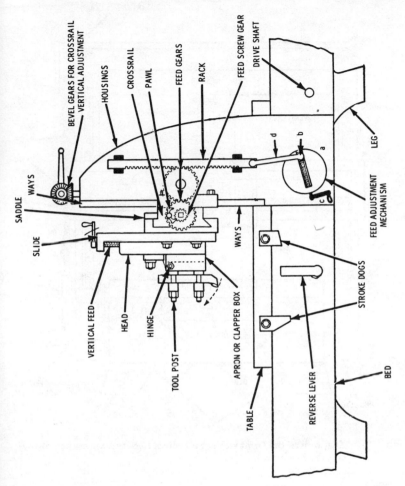

Fig. 19-1. Diagram of the side view of a typical planer, showing the basic parts.

the table, because the accuracy of the surfaces produced is dependent on the accuracy at which the cutting tool is moved (Fig. 19-3). This parallelism can be determined by placing an indicator in the tool post with the point of the indicator touching the table; then moving the toolhead crosswise of the machine, observing at the same time any variation of the indicator needle.

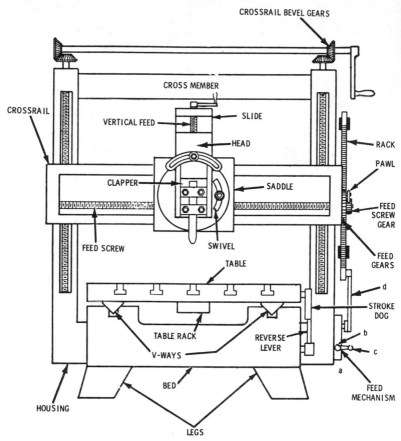

Fig. 19-2. Diagram of the front view of a typical two-column planer, showing the basic parts.

Table

The planer table is a precision-machined plate that travels on
ways of the bed. It presents a broad surface for mounting the
workpiece. The table is provided with T-slots for bolting workpieces and accurately reamed holes for locating stops. Both T-slots
and holes should be kept free from nicks and burrs. The shanks of
the stops should never be forced into the holes by driving them
with a hammer—this action upsets the surface of the table and
destroys its accuracy.

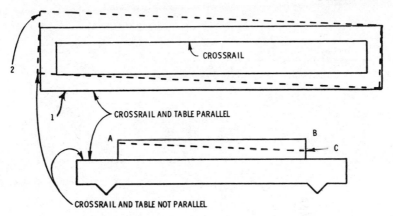

CROSSRAIL

2

1 CROSSRAIL AND TABLE PARALLEL

A   B

C

CROSSRAIL AND TABLE NOT PARALLEL

Fig. 19-3. Diagram showing the lack of parallelism of the crossrail and table.

Toolhead

The toolhead of a planer is similar to that of a shaper both in construction and in operation. A feed screw is provided to move the toolhead with respect to the work. The toolhead can be swiveled for taking angular cuts, and it can be set over in either direction to provide tool clearance when taking vertical or angular cuts on the planer.

Table Drive

The table drive of a planer can be obtained by one of three methods: (1) rack and spur gears, (2) spiral rack and worm, and (3) crank. A basic diagram illustrating the rack-and-spur-gear method with belt-pulley drive is shown in Fig. 19-4.

When the drive mechanism is made up entirely of gears, a large gear known as a bull gear is connected with a rack on the bottom side of the table and to an electric motor by a series of gears. The quick return of the planer table is accomplished by adjustable stops on the side of the table, which, at the end of each cutting stroke, come in contact with a lever which engages a high-speed gear in the driving train of gears.

In the belt-driven type of planer, two belts, one open and one crossed, operate on loose and fixed pulleys. The table travel stops come in contact with a shifter lever at the end of each stroke. As it

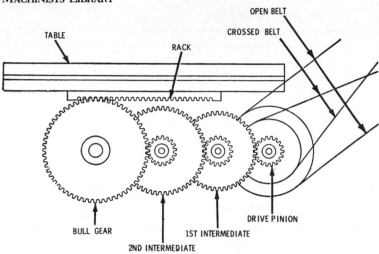

Fig. 19-4. Diagram of rack-and-spur-gear table drive mechanism.

makes contact at the end of the cutting stroke, the lever throws the belt off the fixed pulley onto the loose pulley; at the same time, it throws the crossed belt from a loose pulley to a fixed pulley. As the crossed belt is running faster than the open belt, the table moves faster on the return stroke. The operator can shift the lever by hand to run the belts on loose pulleys, thus stopping the movement of the table without stopping the whole machine. The lever can be locked in neutral position to prevent accidental starting of the machine.

The open and crossed-belt drive mechanism permits operation of the gear train in such a manner that the table will travel slowly on the cutting stroke, reverse, and travel faster on the return stroke. Four pulleys (two larger pulleys and two smaller pulleys) are required (Fig. 19-5). One of the larger pulleys and one of the smaller pulleys are keyed to the drive pinion shaft and are called tight pulleys to distinguish them from the other two pulleys, which turn freely on the shaft and are called loose pulleys.

The larger tight pulley is used for the slower forward speed, or cutting-stroke drive, and the smaller tight pulley is used for the quicker return stroke. Both belts are driven by wide-faced pulleys of the same diameter placed on the countershaft. The open belt on the larger tight pulley drives the machine at a slower speed than

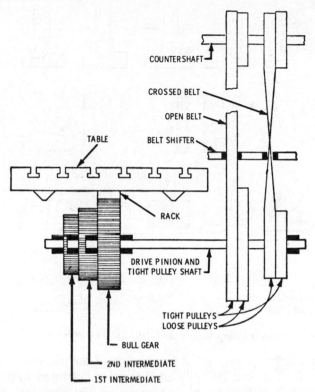

Fig. 19-5. Basic diagram of open and crossed-belt drive of a planer table drive mechanism.

the crossed belt on the smaller tight pulley. In actual operation, both belts run continually and can be shifted back and forth by the belt shifter, which is linked to the reverse lever.

PLANER TOOLS

Cutting tools used on the planer are similar to those used on the lathe (Fig. 19-6). However, the planer is used to machine flat surfaces, while circular surfaces are machined in lathe operations. The chief difference in cutting principles is that the cutting tool on a lathe tends to spring away from the work when it is set at exact center height, but the cutting tool on a planer tends to dig into the

393

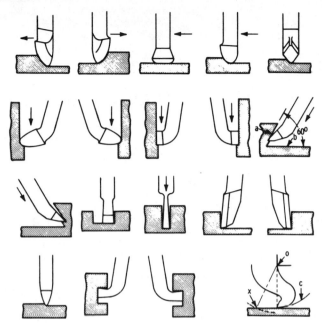

Fig. 19-6. Kinds of cutting tools commonly used on planers.

work if its cutting edge is set in advance of the plane of support (Fig. 19-7). This can be avoided if the tool is forged so that the cutting edge is behind the plane of support. The characteristics of the workpiece determine the number and variety of cutting tools required.

PLANING OPERATIONS

A variety of clamps is available for holding the workpiece on the planer table. Clamping involves the use of such items as bolts, studs, washers, shims, nuts, step blocks, toe dogs, stops, strap clamps, and C-clamps.

Planing Horizontal Surfaces

A typical example of work that can be machined horizontally is the flanged cast-iron cover (Fig. 19-8). This kind of casting is not difficult to clamp to the table. Mount the cast-iron cover with its

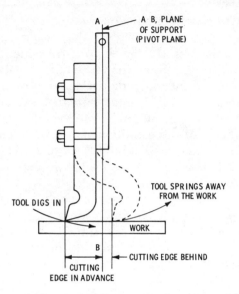

Fig. 19-7. The reason for placing the cutting edge of the planer tool behind the plane or support is to prevent "digging in" of the tool.

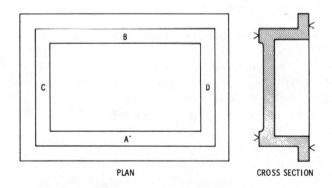

Fig. 19-8. Diagram (left) and sectional view (right) of a flanged cover as a typical workpiece for showing the planing horizontal surfaces.

flanged side downward on the planer table, adjust the crossrail to the correct height, and set up the clamp screws.

The roughing tool should be placed in the toolholder with the cutting tool perpendicular to the work. The tool should be placed

395

against the two clamping bolts on the side so that lateral thrust cannot cause the tool to shift. Tighten the cutting tool bolts so that the clamps are parallel with the clapper—not tilted.

Position and set the stroke for machining surface A (see Fig. 19-8). The depth of the roughing cut depends on the amount of metal to be removed. A deep cut cannot be taken on hard metals because the metal will break and leave a ragged edge at the end of the stroke. Check for this condition at the end of the first stroke; and if the metal tears, set the tool for a lighter cut and take two roughing cuts if necessary. Machine until the cutting tool reaches the inside edge of surface A.

Then position and set the stroke to machine surface D to the inside edge of surface B. Similarly, after machining surface D, reset and position the stroke for surface B. Resetting can be avoided on small workpieces by planing the entire surface at one setting of the cutting tool.

The metal should be roughed until about 0.015 inch of metal is left for finishing. Prepare the work for finishing by breaking the front edge of the surface (the edge at which the tool begins to cut) with a coarse double-cut file. This is done to keep the scale on the outside surface of the casting from destroying the fine cutting edge of the finishing tool.

In taking the finishing cut, extreme care should be taken to finish to the dimension of the blueprint. It should be remembered that if the first cut does not remove enough metal, another light cut can be taken; if the first cut removes too much metal, the work can be ruined.

When the casting is turned over to machine the second side, the casting is automatically leveled, because the machined surface will be in contact with the table surface; therefore, the two sides should be parallel after both sides have been machined. Of course, different clamping methods are necessary to hold the work for machining the second side.

Thin castings should be rough planed on all sides first and then finished. Certain conflicting stresses are set up in the piece when it is cast, because the outer surface becomes chilled first. This places a stress on the scale or skin of the casting; removal of this scale allows the metal to adjust itself to these internal stresses, and the surface will not be true when it is planed.

Planing at an Angle

Planing must be performed at a given angle for dovetails, V-shaped grooves, etc. The toolhead assembly must be pivoted with respect to the saddle so that it can swing around a center axis.

Graduations on the circular part indicate the angular settings in degrees. However, it cannot be assumed that the correct angle is indicated when the head is set at a given angle on the scale—especially on old and worn planers—because more or less lost motion can be present. Therefore, after setting by scale, the setting should be checked as shown in Fig. 19-9.

Mount a dial indicator in the toolholder, clamp a protractor to the table, and set the head at the desired angle (see Fig. 19-9). Adjust the head until the dial indicator contacts the protractor (position A). Feed downward until the indicator reaches position B, and note the reading. If the two readings are the same, the setting is correct.

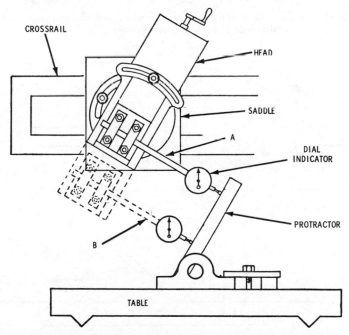

Fig. 19-9. A method of using a protractor and dial indicator to check the angular setting of the slide head.

The planer table V-block (Fig. 19-10) can be used as an example for planing at an angle. As shown in the end view, the sides of the V-block are at 45° from the vertical axis. Set the toolhead at 45° on the scale, as shown in Fig. 19-11. Clamp the head in position by the

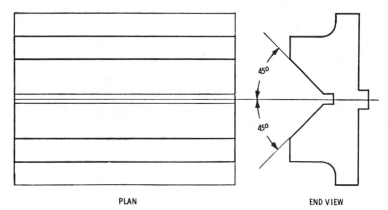

PLAN   END VIEW

Fig. 19-10. Diagram (left) of a planer table V-block with the end view (right) as an illustration of planing at an angle.

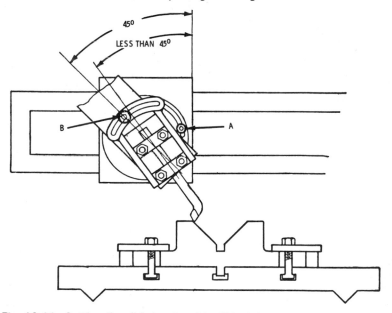

Fig. 19-11. Setting the slide head and tool block for planing the V-block.

bolt *A*. Set the tool block at an angle less than 45°, and clamp in position by tightening the bolt *B*. The tool block is set at less than 45° to prevent the cutting tool dragging over the planed surface on the return stroke (Fig. 19-12). By setting the tool block in the angular position, the cutting edge of the tool swings forward in the plane *YY* at right angles to the tool swing axis *XX* from the work. The angular position of the cutting tool does not affect the direction of tool travel.

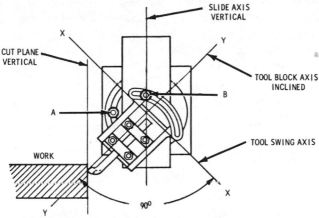

Fig. 19-12. Diagram showing the reason for turning the top of the tool block "away from" the surface to be planed when planing either vertical or inclined surfaces.

The saddle should be moved into position for the first roughing cut (see Fig. 19-11) on the V-block (it is assumed that the bottom side, with its tongue, has been machined). The first cut should be taken at the top. Start the cut with the hand feed, moving downward. Then engage the slide feed, causing the slide and tool to feed downward for each stroke—the saddle remaining in the same position on the crossrail.

After completing the roughing and finishing cuts on one side, do not change the angular setting of the slide for the outer side of the vee; reverse the V-block (turn it 180°) in the table slot, and machine the other side of the vee. This is done because it is impossible to reset the slide to exactly the same angle. Moreover, reversing the work not only ensures machining at the same angle, but the vee will also be in alignment with the center axis of the tongue.

Planing Curved Surfaces

A fixture consisting of a radius arm pivoted on a bracket can be used to plane a concave surface (Fig. 19-13). The feed screw of the slide is removed, and the slide is fastened to the radius arm.

In planing, the cross feed causes the saddle to traverse the crossrail as the tool, which is guided by the radius arm, planes a curved surface. The height of the cutting edge of the tool, as the saddle traverses the crossrail, is determined by the angular position of the radius arm (positions A and B in Fig. 19-13); the locus of all these positions is indicated by the arc.

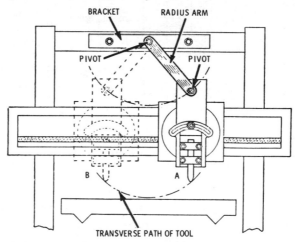

Fig. 19-13. Diagram of a fixture that can be used to plane a circular surface.

Planing a Helix (Work on Centers)

A helix that has a long pitch cannot be cut on some milling machines, but it can be produced on a planer by mounting the work between planer centers and using the fixture, as shown in Fig. 19-14. The fixture consists of a weighted clamp bar and an inclined bar. The upper end A of the inclined bar is attached to the housing, and the lower end B of the inclined bar is attached to the planer bed.

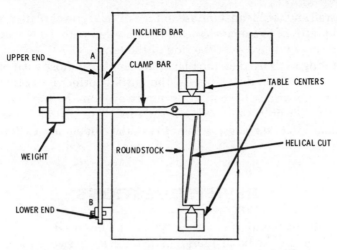

Fig. 19-14. Diagram of a fixture that can be used to cut a helix.

The pitch of the helix depends on the inclination of the bar. The cutting tool is formed to produce the desired helical groove. As the table moves with the clamp bar near the lower end of the inclined bar, the clamp bar, being guided as to angular position, gradually rises as the work turns clockwise through a small arc. Thus, the cutting tool is caused to cut a helix, as shown in Fig. 19-14.

SUMMARY

The planer is a machine that is used on flat or plane surfaces that are fastened to a reciprocating table. The planer differs from the shaper in that the worktable moves back and forth with a reciprocating motion while the cutting tool is held stationary. The planer is used for the same purpose as the shaper, except the planer can handle much larger work and heavier cuts can be taken. The planer is used on work that is too large or otherwise impossible to machine on a shaper.

The most common type of planer is the double-column machine. These columns support the crossrail and house the elevating screws and controls. Another type of planer is the open-side planer, which has only one column or housing to support the

401

crossrail and toolhead. One advantage of this type of planer is that workpieces of irregular shape can be handled with the workpiece extending outward over the side of the table.

Cutting tools used on the planer are similar to those used on the lathe. The chief difference in the cutting principle is that the cutting tool on a lathe tends to spring away from the work, while it tends to dig into the work on a planer. The planer is used to machine flat surfaces, while the lathe machines cylindrical surfaces.

REVIEW QUESTIONS

1. What is the most common type of planer used?
2. What are the basic operation differences between the planer and the shaper?
3. What are the two types of planers manufactured?
4. What is the advantage of the planer over the shaper?

CHAPTER 20

The Slotter

A slotter or slotting machine is, in many respects, a heavy-duty vertical shaper. The chief difference between the slotter and the shaper is that the ram moves in a vertical direction on the slotter.

The slotter can be used for a variety of work other than slotting; although its original purpose has been changed, the machine is still referred to as a slotter. In addition to slotting work, several other kinds of workpieces can be machined more advantageously by a tool that cuts in a vertical direction. Regular and irregular surfaces, both internal and external, can be machined on the slotter; and it is especially adaptable for handling large and heavy pieces of work that cannot be handled easily on other machines.

BASIC PARTS

A diagram of a typical crank-driven slotter is shown in Fig. 20-1. The *ram* carries the cutting tool and has a reciprocating motion. It is similar to the ram of a shaper; but it is more massive and moves vertically, or at right angles to the table, instead of having the horizontal motion of a shaper. A connecting rod links the crank with the ram, and changes the rotary motion of the crank to reciprocating motion, which causes the ram to move up and down. A stroke adjusting screw can be turned to move the crank either toward or away from the crank disk center, which either shortens or lengthens the stroke of the ram.

The *table* of a slotter can be moved either transversely or longitudinally, or it can be rotated about its center. The *hand-feed mechanism* consists of three screw rods, operated by a crank handle, which can be attached to the squared ends of the feed rods. The positions for the crank handle are indicated by the letters (*A*, *B*, and *C* in Fig. 20-1) for operating the transverse, longitudinal, and circular feeds respectively.

The *power-feed mechanism* is, of course, more complicated than the hand-feed mechanism. It consists of a reciprocating-oscillating unit and a transmission gearing unit. The feed mechanisms of various machines differ in detail from the basic diagram (see Fig. 20-1), but the basic principles of operation are similar for all machines.

SLOTTER OPERATIONS

Workpieces that are to be machined to the bottom side cannot be mounted directly on the table surface, but should be mounted on parallel strips to give clearance for the cutting tool (Fig. 20-2). The height *A* of the parallel should be a large enough distance to provide work clearance *B* and ample table clearance *C* for the cutting tool at the end of its cutting stroke. A broken cutting tool and serious damage to the machine can result from insufficient table clearance.

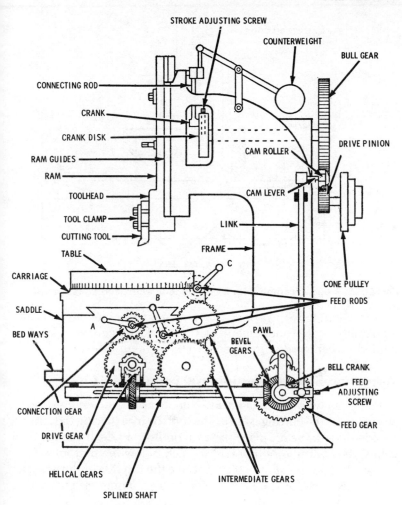

STROKE ADJUSTING SCREW

COUNTERWEIGHT

BULL GEAR

CONNECTING ROD

CRANK

CRANK DISK

RAM GUIDES

RAM

TOOLHEAD

TOOL CLAMP

CUTTING TOOL

TABLE

CARRIAGE

SADDLE

BED WAYS

CONNECTION GEAR

DRIVE GEAR

HELICAL GEARS

SPLINED SHAFT

CAM ROLLER

DRIVE PINION

CAM LEVER

LINK

FRAME

CONE PULLEY

FEED RODS

PAWL

BEVEL GEARS

BELL CRANK

FEED ADJUSTING SCREW

FEED GEAR

INTERMEDIATE GEARS

A   B   C

Fig. 20-1. Diagram of the basic parts layout of a crank-driven slotter.

Clamping the Work

The same clamping methods used for shapers and planers can be used on slotters. Although the ram has a vertical stroke instead of a horizontal stroke (as in the shaper and planer), the cutting stroke of the tool exerts a side pressure on the workpiece. This can

405

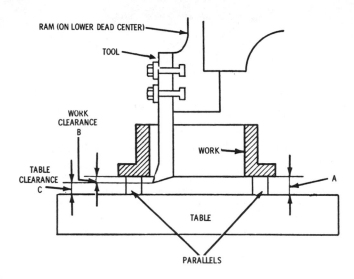

Fig. 20-2. Proper clearances for a cutting tool, work, and table.

be counteracted by proper placement of stops to prevent any possibility of the movement of the work. Circular workpieces should be centered with respect to the center of the table.

Adjusting Stroke Position and Length of Stroke

Place the workpiece in parallels in its approximate position to barely clear the cutting edge of the tool. Before placing the cutting tool in the ram head, place the ram on bottom dead center; then adjust the cutting tool to travel ¼ inch below the bottom side of the work. The tool should travel above the top side of the work an ample distance for the power feed to operate before the cutting tool begins the next cut.

On some machines the stroke adjusting screw is inconveniently reached, when the crank is "on center." If this occurs, the crank disk can be given a quarter-turn, so that the adjusting screw can be turned. Then the stroke can be estimated "by eye," and readjusted until a stroke of proper length is obtained. A stroke indicator with a graduated scale is available on some machines so that a stroke of proper length can be obtained with one setting—avoiding the trial-and-error method.

Alignment of the Work

After the work to be machined has been carefully laid out with clear and distinct scriber lines, place a scriber in the ram head so that its point barely clears the work. Adjust the table position with the traverse hand feed (A in Fig. 20-1) so that the scriber is opposite one end of the line scribed on the work. Then bring the scribed line directly below the scriber tool by turning the longitudinal hand feed B.

Check the setting of the work with the traverse hand feed. As the work moves, the scribed line on the work should follow the scriber mounted in the ram head if the work has been aligned properly. Otherwise, adjustment of the circular position is necessary. To adjust the circular position of the table, turn the circular feed rod handle C, and check as before. Repeat angular adjustment until the work is properly aligned transversely.

Straight, or Flat, Slotter Work

A rectangular open box-like piece of work can be used to illustrate straight, or flat, machining on the slotter (Fig. 20-3). The four sides of the box are to be machined externally to the scribed lines.

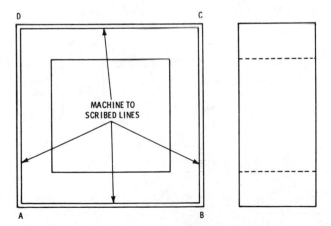

Fig. 20-3. Front view (left) and side view (right) of an open box-like casting, used to illustrate straight machining on the slotter. Note the scribed layout lines for indicating finishing limits.

407

Select parallels, place the work on the parallels, and clamp securely to the table. If the stroke position and length are adjusted properly, the cutting tool should be in the position shown in Fig. 20-4 when the ram is on bottom dead center. Extreme care should be taken to make certain that the work clearance and table clearance are large enough for the cutting tool to clear the table on bottom dead center.

The cut should be started at one end for the side *AB* (see Fig. 20-4) and the entire side machined by vertical downward strokes combined with the traverse feed. Similarly, the opposite side *DC* should be machined without disturbing the mounting of the workpiece on the table—by raising the ram position to top dead center and turning the cutting tool 180°. That is, the cutting tool should be in position *1* for machining the *AB*, and in position *2* for machining the opposite side *DC* (see Fig. 20-4). To machine the second side, the tool can be brought into position by means of the longitudinal feed screw and fed by the traverse feed.

For the other two sides (*BC* and *AD*), turn the table 90° by means of the circular feed rod (*C* in Fig. 20-1), and proceed to

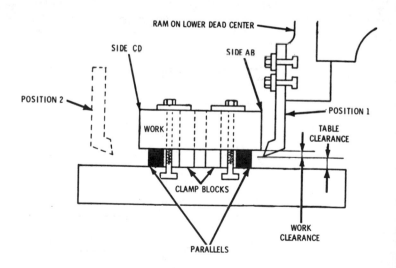

Fig. 20-4. Setup for machining the open box-like casting. The dotted lines (tool position 2) indicate the position of the cutting tool after reversal.

machine the sides as before. On some slotters, the ram head which carries the cutting tool can be swiveled to any angular position. Four graduations are provided (90° apart). Thus, all four sides of the work can be machined by swiveling the tool—making it unnecessary to reverse the cutting tool or to change the angular setting of the table. For precision machining, the scribed lines merely serve as a guide; the final finishing cut should be made to precise caliper measurements.

Circular Slotter Work

The slotter is well suited for machining cylindrical surfaces (Fig. 20-5). The work must be properly aligned, that is, centered with the center of the table. In other words, the work must be positioned on the table so that the axis of the cylindrical surface to be machined coincides with the axis of rotation of the table. The circular feed screw (C in Fig. 20-1) rotates the table to produce a small arc after each cutting stroke of the tool.

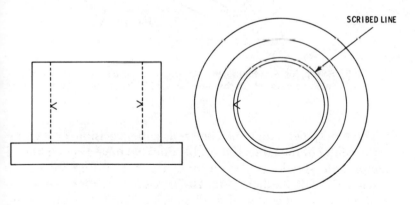

SCRIBED LINE

Fig. 20-5. Front view (left) and top view (right) of a flanged casting that is typical of circular slotter work.

The center hole of the table is used as a guide for centering small workpieces; some machines have concentric circles cut in the table for centering large workpieces (Fig. 20-6). The setup for machining a simple flanged cylinder (see Fig. 20-5) is illustrated in Fig. 20-6.

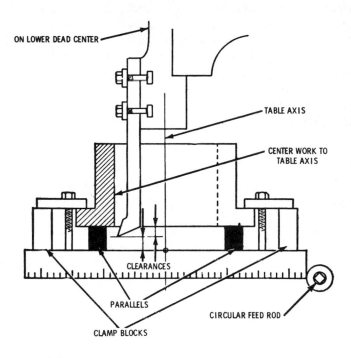

Fig. 20-6. Setup for machining the flanged cylindrical casting. Note the table axis on which the work should be centered.

The workpiece should be centered, with precision, concentric to the center of the table. If the table does not have a center hole, a center hole should be bored, and a center fitting should be machined to fit the hole (Fig. 20-7). A center fitting should be included as an accessory for all slotters. The fitting should be machined accurately and stored with the same care that should be given calipers and other delicate measuring equipment.

To center the workpiece, clamp a scriber in the ram head, so that its point will barely clear the top of the center fitting. Shift the position of the table by means of the transverse and longitudinal feeds until the point of the scriber is directly over the center of the fitting. After centering the scriber tool, do not change the longitudinal setting of the table.

410

Raise the ram to upper center to place the scriber out of the way. Place the workpiece on parallels on the table in an approximate concentric position "by eye," and true up with hermaphrodite calipers (Fig. 20-8). Set the calipers to one-half the bore diameter, making allowance for thickness of the metal to be removed.

In truing up with the hermaphrodite calipers, caliper two radii on the same diameter, shifting the work along this diameter until centered. Repeat on a diameter at 90° (located by eye). Clamp the work lightly, remembering not to change the longitudinal setting of the table; move the table over with the transverse feed, and lower the ram until the point of the scriber previously clamped in the ram head is directly over and barely clears the point (A in Fig. 20-8) on the scribed layout circle.

Remove the scriber, and clamp the roughing tool in position (see Fig. 20-2). Raise the arm to bring the cutting tool slightly above the top surface of the work. Shift the table with the transverse feed to bring the tool to proper axial position (A in Fig. 20-9). Clamp the carriage to the saddle to prevent any transverse movement, and start the machine.

After two or three strokes of the cutting tool, check to make certain that the stroke and cutting tool are adjusted properly. Then feed the cutting tool into the work to the proper depth for a roughing cut by means of the longitudinal feed, and lock the saddle to prevent any change in the setting of the longitudinal feed. Take the cut, using the circular feed; rough cut to near the layout circle, leaving only enough metal for the finishing cut or cuts.

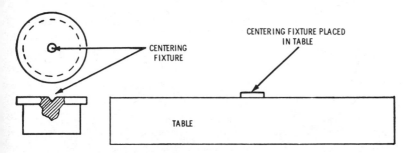

Fig. 20-7. Diagram of a centering fixture (left), and its application as a table accessory (right) for the slotter.

411

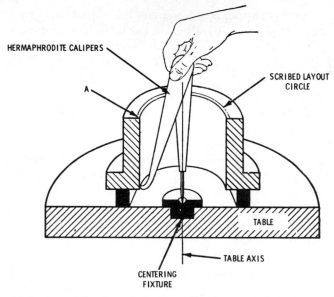

Fig. 20-8. An application of the centering fixture and the hermophrodite calipers for centering circular work. Note the scribed layout circle that can be used as a reference guide for the final centering check.

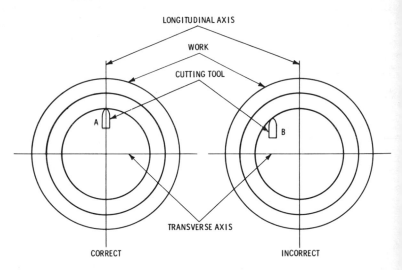

Fig. 20-9. The correct and incorrect positions for the slotter cutting tool.

SLOTTER CUTTING TOOLS

As the cutting motion of a slotter is vertical, rather than horizontal, slotter cutting tools differ from shaper and planer tools. The cutting edge is formed on the end of the cutting tool so that it is placed under compression and will cut when pushed endwise. The cutting face that turns the shaving is on the end of the tool (Fig. 20-10).

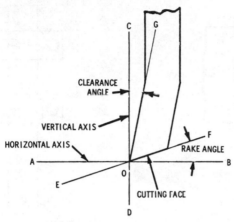

Fig. 20-10. The correct clearance and rake angles for the slotter cutting tool.

The horizontal axis *AB* is parallel with the table, and perpendicular to the vertical axis *CD*. The clearance angle *COG* should be 4 or 5 degrees, and the rake angle *BOF* should be 10 or 12 degrees. When the cutting tool is clamped to the end of the ram in a horizontal position, the clearance and rake angles are measured in the same manner as for planer tools.

A typical set of slotter cutting tools includes the following: (1) roughing; (2) finishing; (3) right hand; (4) left hand; (5) keyway; and (6) scriber. The pointed scriber is a much-used tool for cylindrical work setups. In addition to the aforementioned tools, several special tools are designed for certain kinds of work. For example, a heavy slotter bar with an inserted cutting tool is better adapted for external cuts than tools that are forged in a single piece. Properly held in a round bar that can be turned around, the inserted cutting tool can be set at any angle (Fig. 20-11). This adapts the tool for machining slots or keyways, using a broad-nosed cutter of desired shape and width. For cutting square holes, the bar can be turned to reach the corners. Cutting tools inserted in a bar should not be given too much rake because the tool drags over the work on the upward stroke.

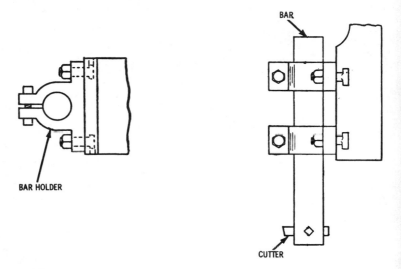

Fig. 20-11. Detail of the holder for a round bar (left) and the round bar and inserted cutting tool (right). Note that the cutting tool is in a horizontal, rather than a vertical, position.

FEEDS AND SPEEDS

Longitudinal, transverse, and circular feeds are provided on the slotter as both hand and power feeds. Rapid power traverse is provided on most slotters, particularly the larger sizes of slotters. They are constructed so that the feed and rapid power traverse are interlocking; therefore, it is impossible to engage both at the same time. Approximate cutting speeds for various metals are given in Table 20-1.

Table 20-1. Slotter Cutting Speeds (Feet per Minute)

MATERIAL	ROUGHING		FINISHING	
	C.S. Tools	H.S. Tools	C.S. Tools	H.S. Tools
Aluminum	50	125	40	60
Babbitt	50	Highest	30	80
Brass	35	Highest	30	80
Cast iron (hard)	35	20	10	20
Cast iron (soft and medium)	30	60	15	45
Copper	40	80	20	30
Monel metal	40	80	20	30
Steel (hard)	15	35	10	25
Steel (soft)	35	Highest	25	50

SUMMARY

A slotter, in many respects, is a heavy-duty vertical shaper. The slotter can be used for a variety of work other than slotting. In addition to slotting work, several other kinds of pieces can be machined more advantageously by a tool that cuts in a vertical direction. Regular and irregular surfaces, both internal and external, can be machined. The slotter is especially adaptable for handling large and heavy pieces of work that cannot be handled easily on other machines.

A typical set of slotter tools includes the roughing, finishing, keyway, and scriber. The scriber is used many times for cylindrical work setups. In addition to these tools, several special tools are designed for certain kinds of work. Heavy slotter bars with an inserted cutting tool are better adapted for external cuts than tools that are forged in a single piece.

REVIEW QUESTIONS

1. What is the difference between a slotter and a shaper?
2. What are the advantages in using a slotter machine?
3. What is the ram?

CHAPTER 21

Abrasive Metal Finishing Machines

There are two major types of grinding: nonprecision and precision grinding. Nonprecision grinding is also called off-hand grinding. Metal is removed by this method when there is no great need for accuracy. Pedestal or bench grinders are used for this type of grinding; see Figs. 21-1 and 21-2. These grinders are also used for rough grinding and sharpening tools.

In precision grinding, metal can be removed with great accuracy. There are a number of different precision grinding machines available that can grind metal parts to different shapes and sizes with very accurate dimensions; see Figs. 21-3 and 21-4.

The grinding wheels used in both nonprecision and precision grinding are made from aluminum oxide or silicon carbide. These grinding wheels come in many different shapes, faces, and sizes. Manufacturers of grinding wheels have standardized these shapes and faces; see Figs. 21-5 and 21-6. The wheels are made in a wide range of sizes for most grinding jobs. A wide variety of mounted grinding wheels are also made for use with small grinders for either off-hand or precision grinding on dies; see Fig. 21-7.

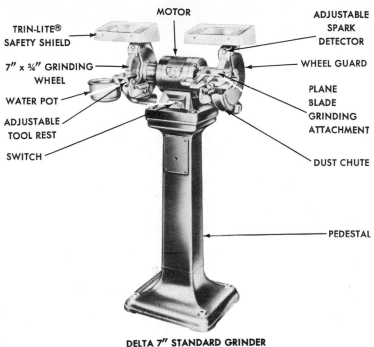

MOTOR

TRIN-LITE® SAFETY SHIELD

ADJUSTABLE SPARK DETECTOR

7" x ¾" GRINDING WHEEL

WHEEL GUARD

WATER POT

PLANE BLADE GRINDING ATTACHMENT

ADJUSTABLE TOOL REST

SWITCH

DUST CHUTE

PEDESTAL

DELTA 7" STANDARD GRINDER

Courtesy Rockwell Manufacturing Co.

Fig. 21-1. Pedestal grinder.

Grinding wheels should be properly mounted on grinders because they are operated at very high speeds; see Fig. 21-8. When the wheels become loaded with particles of material from a grinding operation, they must be dressed. Dressing restores the grinding wheel to its original shape with a clean cutting face.

The three types of grinding wheel dressers in use for precision grinding machines are: the mechanical dresser, the abrasive wheel, and the diamond tool. The diamond tool is the most commonly used wheel dresser on precision grinders; see Fig. 21-9.

Grinding wheels vary by:

1. Type of abrasive. Aluminum oxide or silicon carbide.
2. Grain sizes. Grain size in wheels range from 600 (fine) to 10 (coarse).

Courtesy Rockwell Manufacturing Co.

Fig. 21-2. A 6-inch edge tool bench grinder.

3. Type of bond. Material that holds the wheel together.
4. Structure. Structure or grain graining in a wheel is designated by numbers from 1 to 12. The lower the numbers, the closer the grain spacing; see Fig. 21-10.
5. Grade. Grades range from A to Z (soft to hard). Grades A thru H are soft, grades I thru P are medium, and grades Q thru Z are hard.

The standard grinding-wheel marking system is based upon these factors; see Fig. 21-11.

There are many different grinding methods used. Surface grinders are used in the grinding of precision parts; see Figs. 21-12 and 21-13. They come in many different shapes to perform the wide variety of grinding operations required in industry.

Abrasive Blasting

Abrasive blasting is used when a quality surface finish is not of primary importance. Abrasive material is blasted on metal surfaces with great force. Usually abrasive blasting takes place in a blasting room, on tables, or in blasting cabinets. The use of respirators is required for blasting in blasting rooms.

419

Fig. 21-3. Form-relief grinder. A single machine for complete form-relief grinding and precision sharpening of a large variety of cutting tools.

Barrel and Vibratory Tanks

One of the most popular methods of finishing parts is called barrel finishing. Barrel finishing is also known as abrasive tumbling. Abrasive tumbling is a precision-controlled method of re-

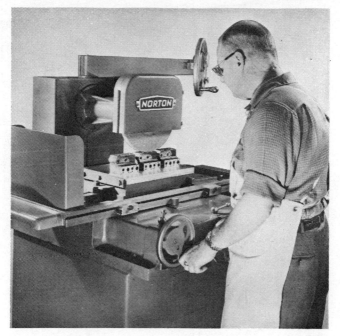

Fig. 21-4. A hand operated surface grinder.

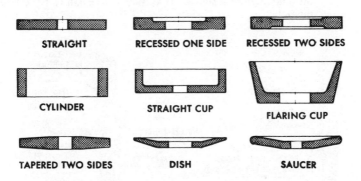

STRAIGHT   RECESSED ONE SIDE   RECESSED TWO SIDES

CYLINDER   STRAIGHT CUP   FLARING CUP

TAPERED TWO SIDES   DISH   SAUCER

Fig. 21-5. Nine standard shapes for grinding wheels. These shapes will perform many jobs.

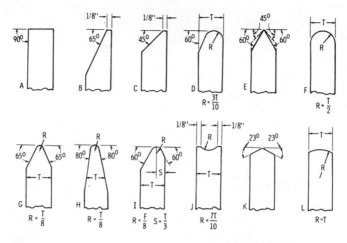

Courtesy Cincinnati Milacron Co.

Fig. 21-6. Standardized grinding-wheel faces. These faces can be modified by dressing to suit the needs of the user.

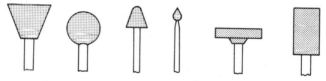

Courtesy Cincinnati Milacron Co.

Fig. 21-7. Mounted points are tiny grinding wheels permanently mounted on small-diameter shanks. They may be as small as 1/16 inch in diameter.

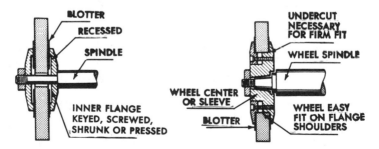

Courtesy Cincinnati Milacron Co.

Fig. 21-8. Correct method of mounting a grinding wheel. Never omit the blotting-paper washers.

Fig. 21-9. Diamond truing tools must be canted, as shown, to prevent chatter and gouging of the wheel and to maintain sharpness of the diamond abrasive.

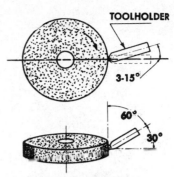

TOOLHOLDER

3-15°

60°

30°

Courtesy Cincinnati Milacron Co.

Courtesy Norton Company

Fig. 21-10. Grinding-wheel structures showing medium (left) and wide (right) grain spacings.

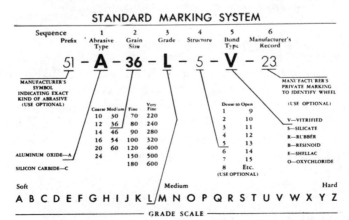

STANDARD MARKING SYSTEM

Sequence	1	2	3	4	5	6
Prefix	Abrasive Type	Grain Size	Grade	Structure	Bond Type	Manufacturer's Record

51 – A – 36 – L – 5 – V – 23

MANUFACTURER'S SYMBOL INDICATING EXACT KIND OF ABRASIVE (USE OPTIONAL)

MANUFACTURER'S PRIVATE MARKING TO IDENTIFY WHEEL (USE OPTIONAL)

Coarse	Medium	Fine	Very Fine
10	30	70	220
12	36	80	240
14	46	90	280
16	54	100	320
20	60	120	400
24		150	500
		180	600

ALUMINUM OXIDE—A

SILICON CARBIDE—C

Dense to Open
1	9
2	10
3	11
4	12
5	13
6	14
7	15
8	Etc.
(USE OPTIONAL)

V—VITRIFIED
S—SILICATE
R—RUBBER
B—RESINOID
E—SHELLAC
O—OXYCHLORIDE

Soft   Medium   Hard

A B C D E F G H I J K L M N O P Q R S T U V W X Y Z

——————— GRADE SCALE ———————

Courtesy Aluminum Company of America

. Standard grinding-wheel marking system.

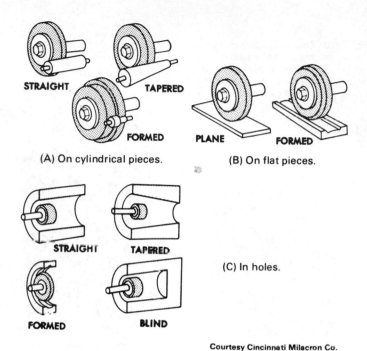

STRAIGHT TAPERED

FORMED PLANE FORMED

(A) On cylindrical pieces. (B) On flat pieces.

STRAIGHT TAPERED

(C) In holes.

FORMED BLIND

Courtesy Cincinnati Milacron Co.

Fig. 21-12. Types of grinding commonly performed on external, internal, and flat work. Centerless grinding, either external or internal, is also performed on cylindrical work.

moving sharp burrs, edges, and heat-treat scale from parts; see Fig. 21-14. This process also improves the surface finish of a part. A number of parts in the many different products that we use in our daily lives are finished with this process.

In the barrel finishing process abrasive media is put into a vibrating or rotating barrel along with water and a chemical compound. The action that takes place within the barrels removes the sharp burrs, edges, and heat-treat scale and then polishes the part; see Fig. 21-15.

Belt Sanders

Metal is also ground with belt sanders. Grinding requires the use of special safety equipment to protect the eyes and hands; see Fig.

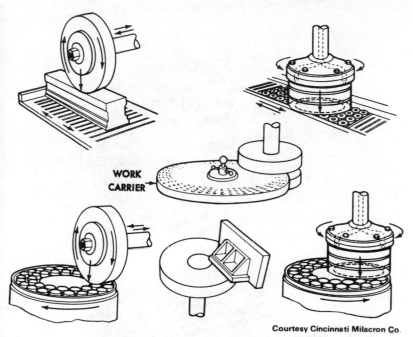

WORK CARRIER

Courtesy Cincinnati Milacron Co.

Fig. 21-13. Principle types of surface grinding include flat and rotary tables, and horizontal and vertical wheels. Disk sanders may be single- or double-wheel and either vertical or horizontal.

21-16. It is a dangerous operation. Safety rules must be observed at all times.

Safety Rules for Abrasive Metal Finishing

Grinding Wheels

1. Check for balance and ruts. Unbalanced and rutted wheels can cause work pieces to be thrown.
2. Operate at the proper speed. Too high an operating speed can cause the wheel to burst.
3. Check for worn center holes. Worn center holes can cause the work piece to be thrown.

Abrasive Blasting

1. Check for secure hose connection. Accidental disconnection of nozzle from hose can cause hose to whip.

425

Courtesy Aluminum Company of America

Fig. 21-14. A typical part before (top) and after (bottom) vibratory barrel finishing.

2. Check for "dead man" switch. No "dead man" switch to turn off blasting if hose is dropped could cause injury.
3. Check for clear glass in blasting helmet window. Poor visibility through blasting helmet window due to dirty or scratched glass could give serious problems to the operator.
4. Check for excessive hose wear. Excessive wear of hose can cause hose to break.

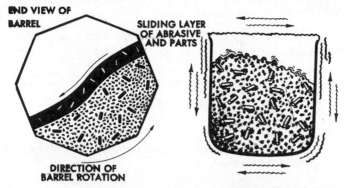

END VIEW OF BARREL

SLIDING LAYER OF ABRASIVE AND PARTS

DIRECTION OF BARREL ROTATION

Courtesy Norton Company

Fig. 21-15. Barrel-finishing action within a rotating barrel (left) and a vibratory barrel (right).

Courtesy Aluminum Company of America

Fig. 21-16. An operator of a belt sander finishing an aluminum casting.

Barrel and Vibratory Tanks

1. Check for a lockout device. Barrels and tanks can be accidentally turned on during loading unless they have a lockout device.
2. Check for unguarded pinch points. Barrels and tanks have many pinch points that may not be properly guarded.

Belts

1. Check for cracked and torn belts. Cracked or torn belts can whip around when broken.
2. Check for unguarded belts. Unguarded belts can allow you to come in contact with the belt.
3. Operate at the proper speed. Too high an operating speed can cause belts to break.

SUMMARY

There are two major type of grinding: nonprecision and precision grinding. Nonprecision grinding is also called off-hand grinding. Metal is removed by this method when there is no great need for accuracy. Pedestal and bench grinders are used for this type of grinding.

In precision grinding, metal can be removed with great accuracy. There are a number of different precision machines available that can grind metal parts to different shapes and sizes with very accurate dimensions.

Abrasive blasting is used when a quality surface finish is not of primary importance. Abrasive material is blasted on metal surfaces with great force.

Barrel and vibratory tanks are used for abrasive tumbling. Abrasive tumbling is a precision controlled method of removing sharp edges, burrs, and heat-treat scale from parts. The process also improves the finish of a part.

Metal is also ground with belt sanders. This type of grinding requires the use of special safety equipment to protect the eyes and hands. In fact, all the various types of abrasive metal finishing require that safety rules must be observed at all times.

REVIEW QUESTIONS

1. What are two types of grinding?
2. What is abrasive blasting?
3. What is the purpose of barrel finishing?
4. What types of work are finished on belt sanders?

CHAPTER 22

Electroforming

The development of newer and tougher alloys and other materials has increased the demand for processes and machines that are able to machine these materials more quickly and economically. Many production items that were formerly produced on grinding and milling machines are now produced on electrochemical and electroforming machines.

ELECTROCHEMICAL MACHINING PROCESS

Electrochemical machines are used for drilling, trepanning, and shaping extremely hard and tough materials (Fig. 22-1). These machines may range from a small drill of the bench type to the larger automatic-cycle production machines.

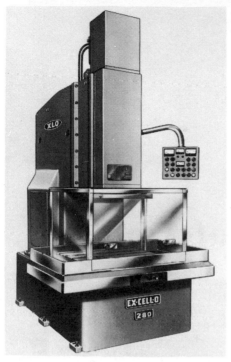

Fig. 22-1. Electrochemical machine.

Courtesy Ex-Cell-O Corporation

Basic Principle

In the electrochemical machine a low-voltage electric current is passed through a conductive fluid between the work and the tool electrode (Fig. 22-2). The fluid also flushes away the residue from the machining operations.

Machine Operation

Some machines may be equipped with push buttons and controls for manual cycling. The machines may also be equipped for either semiautomatic or automatic cycling to suit production requirements.

In the machine shown in Fig. 22-2, a vertical ram supports a platen for mounting the tool electrodes. The platen is insulated

Courtesy Ex-Cell-O Corporation

Fig. 22-2. A conductive fluid carries the electric current that passes between the tool and the work.

from the ram and is connected through bus bars and cables to the electrical power supply. The power supply delivers current to the work area at low voltage.

The electrolyte system consists of a reservoir, a filter, a pump and drive motor, work compartment connections, and a return flow line. Corrosion-resistant materials, such as plastic or stainless steel, are used throughout the electrolyte system. The electrolyte is filtered before it enters the work area.

The type of electrolyte solution used depends on the nature of the materials to be cut and the operations to be performed. Neutral, acidic, or alkaline solutions are used, depending on the operation.

Operations Performed

One advantage of the electrochemical process is that there is no tool wear. Any number of holes may be pierced without wearing away the electrode (Fig. 22-3).

433

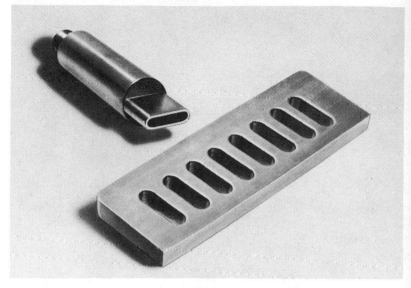

Fig. 22-3. Elongated holes produced by the tool electrode.

Internal gear shapes, as well as round holes and various other shapes, can be produced by the electrochemical process (Fig. 22-4). Tool marks and burrs are completely avoided; therefore, these do not have to be machined off, as in most other processes.

A die, and the electrode used to machine it, are shown in Fig. 22-5. Three-dimensional cavities may be used in the process.

The electrochemical machine can be used as a general purpose production machine. The process can be adapted to drilling, cutting off, shaping, trepanning, broaching, and cavity-sinking operations on any metal that is an electrical conductor.

ELECTRICAL DISCHARGE MACHINING PROCESS

This process can be used to machine hard or tough materials to any form that can be produced in an electrode. It can be used only on electrically conductive materials.

434

Fig. 22-4. A production of internal gears by the electrochemical process.

Fig. 22-5. A die (left) and the electrode (right) used to machine it.

Basic Principle

Electrical discharge machining is performed with both the electrode and the work submerged in a dielectric fluid. The electrode and the workpiece must be in close proximity. Direct-current

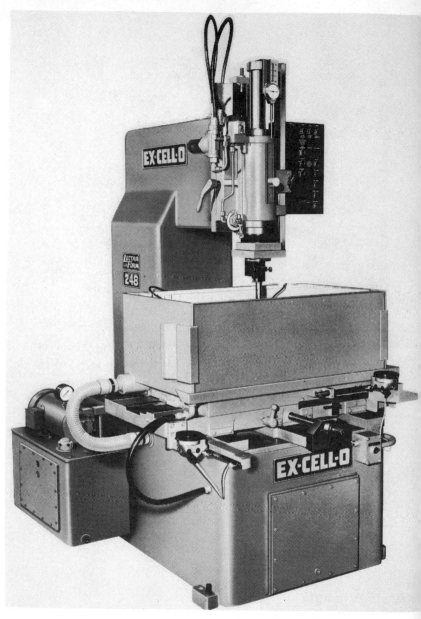

Fig. 22-6. Ex-Cell-O Model 248 Lectro-Form machine. This machine uses the electrical discharge machining (EDM) process of forming metal.

pulses at high frequency are discharged from the electrode through the resistance of the fluid. A minute particle is eroded from the surface of the work by each pulse.

Machine Operation

The electroforming machine shown in Fig. 22-6 is equipped with an electrohydraulic servo for maintaining a constant gap between the electrode and the work. The servo automatically feeds the tool at the proper rate to keep the gap constant as the cutting progresses.

The power supply feeds the current to the servo until the electrode reaches the preset depth, and a predetermined amount of stock is removed. The power supply controls the movement of the electrohydraulic servo unit in the machine tool while precisely metering high frequency direct-current pulses to the electrode. The machine tool is a precision mechanical device. It holds the work and feeds the electrode to the work while maintaining an extremely precise gap. The servo automatically retracts the tool if

Courtesy Ex-Cell-O Corporation

Fig. 22-7. A multiple electrode setup (left) in which 18 holes are drilled in the aircraft fuel nozzle part (right) in one machining cycle.

437

Fig. 22-8. Miscellaneous forms machined by the electrical discharge machining process.

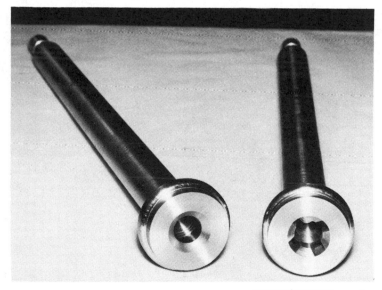

Fig. 22-9. A single-axis movement of the electrode machined this four-lobed spline in the stainless steel shaft.

the gap becomes too small or if residue from the cutting operation forms a bridge. Momentary retraction of the tool clears the cutting area, and the servo resumes its downfeed. Accurate gaging and depth of machining are controlled by the combination of a micrometer and limit switches.

An extremely uniform gap is held between the tool and work, regardless of configuration. Due to the extreme repeatability and uniformity of the process, extremely accurate predictions of results are possible. Size, tolerances, and surface finish of the work can be predicted within very close limits.

Cutting speed is controlled by amperage, or cutting current. Surface finish of the work is controlled by the frequency at which the cutting current pulses occur.

A workpan is provided on the machine for setup of the work. An adjustable level control and an adjustable safety float switch are provided to control the level of the dielectric fluid.

Courtesy Ex-Cell-O Corporation

Fig. 22-10. Component parts of a die, used to produce powdered metal cluster gears, showing internal involute gear form rough and finish machined in the hardened condition.

Operations Performed

Electrical discharge machining is particularly adaptable to general-purpose machining and for die work. Various types of work may be done on these machines.

A multiple-electrode setup is shown in Fig. 22-7, where 18 holes are drilled in an aircraft fuel nozzle in a single machining cycle.

Fig. 22-8 illustrates some of the miscellaneous forms that can be machined by the process. Hardened steels and carbides as well as tough alloys can be machined without difficulty (Fig. 22-9).

Electroforming can be used to greatest advantage in die manufacturing and machining operations involving intricate forms that are difficult to produce by conventional machines. This is particularly advantageous in machining of parts in the hardened condition (Fig. 22-10). Dies can be reworked, and reworking can be performed earlier to maintain a high quality-control level in the production of parts.

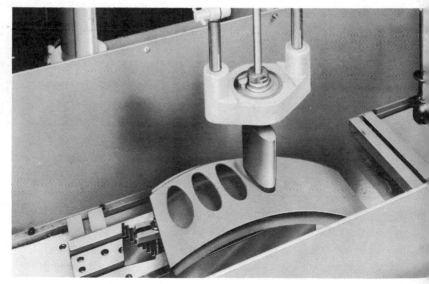

Courtesy Ex-Cell-O Corporation

Fig. 22-11. A typical example of an irregular shape easily cut in tough material by electrical discharge machining. This is a stainless-steel shroud, pierced by a brass electrode positioned above it.

Irregular shapes can be cut easily in tough materials (Fig. 22-11). Forms can be produced with a soft tool. Tool pressure, which can cause localized stress, and heat, which can cause distortion, are both absent in electrical discharge machining.

Fig. 22-12 illustrates other items that may be produced by the electroforming process more quickly than by conventional means. Tools used on conventional machines may be fitted on the electroforming machine, especially after they have been heat treated.

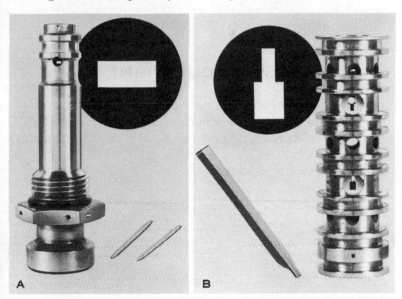

Courtesy Ex-Cell-O Corporation

Fig. 22-12. Valves produced by the electroforming process. (A) A high-pressure relief valve, insert shows the shape of the opening, (B) another valve, (C) shape of the opening is shown in the insert.

SUMMARY

Electrochemical machines are used for drilling, trepanning, and shaping extremely hard and tough materials. In the electrochemical machine, a low-voltage electrical current is passed through a conductive fluid between the work and the tool electrode.

441

One big advantage of the electrochemical process is that there is no tool wear. Many holes can be pierced without wearing away the electrode. Tool marks and burrs are completely avoided; therefore, these do not have to be machined off, as in most other processes.

Electrical discharge machining is performed with both the electrode and the work submerged in a dielectric fluid. Direct-current pulses at high frequency are discharged from the electrode through the resistance of the fluid. Minute particles are eroded from the surface of the work by each pulse.

REVIEW QUESTIONS

1. How are electrochemical machines used in industry?
2. What are the advantages in electrochemical processing?
3. What is the principal of operation of the electrical discharge machine?
4. What is electroforming?

CHAPTER 23

Ultrasonics in Metalworking

Ultrasonics, using a form of cavitation, may be utilized in several different machining processes. Drilling, grinding, cutting, and soldering or welding may all be performed with ultrasonics.

MACHINING

The machine processes use an abrasive slurry, which is pounded against the material at an ultrasonic rate, causing it to be chipped away. As the work is not chipped, heated, stressed, or distorted, a greater degree of precision may be obtained than with conventional machining methods. The size of the abrasive grit used determines the degree of fineness of cut; coarse abrasive grit cuts the material at a faster rate. Ultrasonic machining is especially advantageous in cutting very hard materials such as glass, ceramics, steel, and tungsten; it is practically ineffective on soft substances.

Drilling

An ultrasonic drill concentrates all the vibrations at a specific point, thereby making use of all the available energy at a single point. In concentrating the energy, the amplitude of the vibrations is increased; this is brought about by the use of a tapered stub (Fig. 23-1). In the cross-sectional view it can be noted that the driving coils are wound around the magnetostrictive nickel core. The laminations are firmly connected to the tapered stub, which extends out of the drill housing. As the core changes size, it causes the stub to vibrate up and down to set up longitudinal vibrations, thereby, creating the drilling motion. The assembly has an inlet and outlet for circulation of water to provide cooling, as the vibrations produce heat.

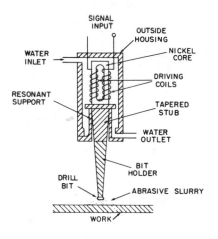

Fig. 23-1. Cross-sectional view of an ultrasonic drill.

The drill tip actually moves up and down only a very short distance, but the rapid rate of vibration makes it very effective in driving through hard materials. This vibrating effect combines with cavitation in the liquid-base abrasive to perform the driving action. The tip of the drill is applied to the abrasive, which performs the actual drilling action. Aluminum oxide, boron carbide, silicon carbide, carborundum, diamond dust, or other similar hard materials mixed with oil or water can be used as the abrasive. A container for the abrasive is mounted so that it can be applied at the point where the drilling is being performed.

Most ultrasonic drilling machines resemble rotary drill presses in appearance. A commercial ultrasonic drill (Fig. 23-2) is arranged so that different bits can be substituted in the same transducer

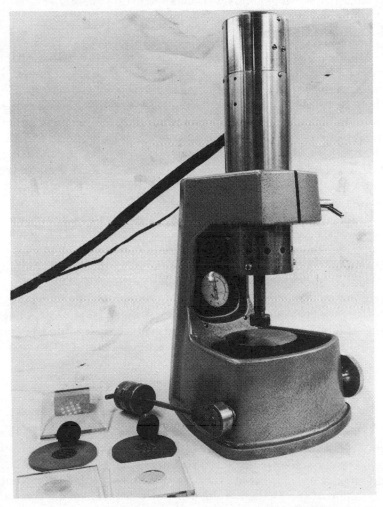

Fig. 23-2. An ultrasonic drill. At the lower left, notice the cuts already made.

assembly. As drilling is by reciprocating action, rather than by rotary motion, each cutting tool is made in the same shape as the hole to be drilled. A rotary drill can drill only round holes, but an ultrasonic drill can cut holes in almost any desired shape. With a properly constructed tool, a large number of holes can be machined in a single operation. Drill bits are made of cold-rolled, or unhardened, steel, then silver-soldered to the core-shaped shaft.

Grinding and Cutting

Grinding and cutting operation by ultrasonics on hard materials are similar to drilling. Grinding involves either shaping a material to a particular form or smoothing it to a particular finish. A coarse abrasive is used at the start; then a finer grain is used until the correct finish is obtained. Ultrasonic grinding has several distinct advantages, as compared with conventional grinding. A good finish may be obtained at a lower temperature, thereby reducing damage. The amount of grinding can be controlled accurately, and a good finish is possible on coarse materials. An ultrasonic grinder is shown in Fig. 23-3.

Ultrasonic grinding and cutting are also advantageous in that they work best with the harder metals, while ordinary machine tools work best with the softer materials. Ultrasonics can be used to cut glass, quartz, sapphire, germanium, silicon, diamond, ceramics, and other hard nonmetallic substances. The various hard materials can be cut or sliced into thin wafers for various uses. With a properly shaped tool, a number of slices can be cut at one time (Fig. 23-4). Cutting blades are usually made of thin steel or molybdenum, and can be used to cut wafers as thin as 0.015 inch.

SOLDERING AND WELDING

When metals such as copper or aluminum are exposed to the atmosphere, a thin film of oxide forms on the metal. Thus, the film prevents the solder from making proper contact with the metal. A chemical flux can be used to coat copper materials and to prevent formation of the oxide, but this will not work well on aluminum. Application of ultrasonic vibrations to molten solder, when in

TRANSDUCER SUPPORTING CYLINDER

GRINDING FORCE ADJUSTMENT

DISPLACEMENT INDICATOR

PARALLELOGRAM SUPPORTED WITH BALL BEARINGS

TOOL ROTATION LOCK

SUPPORT FLANGE

TOOL CONE

COUNTERBALANCING SPRING

Courtesy Raytheon Company

Fig. 23-3. The head unit of an ultrasonic grinder.

contact with aluminum, cavitates the solder and eliminates the film of oxide. Thus, the solder can reach and contact the surface of the aluminum. Copper and other metals can be soldered in the same manner, using solder with a high tin content. After tinning by ultrasonic means, soldering can be done by the usual methods. Ultrasonic soldering usually requires no flux, even for joining dissimilar metals. One type of ultrasonic soldering iron is illustrated in Fig. 23-5.

Ultrasonic welding differs considerably from other types. No

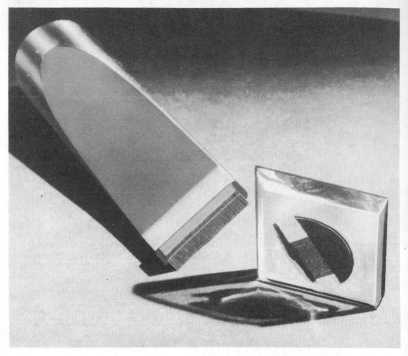

Fig. 23-4. An ultrasonic cutterhead. A large number of slices can be produced with each cut.

external heat is applied—only ultrasonic vibrations—but similar
and dissimilar metals can be joined effectively. Aluminum can be
welded either to aluminum or to another metal. Stainless steel,
molybdenum, titanium, and other metals can be joined either by
spot welding or by seam welding.

The amount of heat created by the ultrasonic vibrations is not
sufficient to melt the metals being joined; therefore, the bond
occurs by some other means—the joining process is not completely understood. Evidently the two sections are joined by a
molecular band formed by the combined action of heat and the
fast rate of vibration. As in other types of welds, the bond is
stronger than either of the metals involved. Special preparation of
materials is not required in ultrasonic welding, and much less
pressure is required to hold the sections together during the pro-

cess. As there is so little heat, there is less deformation of the materials being welded.

One type of ultrasonic welding unit (Fig. 23-6) resembles an ultrasonic drill; it has a tapered stub with a large tip on the end. The vibrations are applied parallel to the surfaces of the materials being welded. These are shear waves—longitudinal waves will not weld. A seam weld is formed by running the tool down strip metals. A spot weld is produced by touching the tool at a definite point.

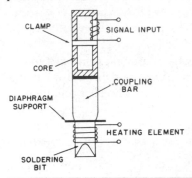

Fig. 23-5. An ultrasonic solder iron.

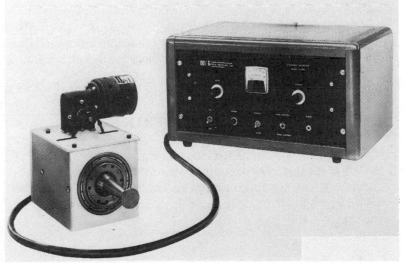

Fig. 23-6. An ultrasonic seam welder. The motor turns the transducer head.

449

SUMMARY

Ultrasonics may be utilized in several different machining processes. Drilling, grinding, cutting and soldering or welding may be performed with ultrasonics. The machine process uses an abrasive slurry, which is pounded against the material at an ultrasonic rate, causing it to be chipped away.

An ultrasonic drill concentrates all the vibrations at a specific point, thereby making use of all the available energy at a single point. The drill tip actually moves up and down only a very short distance, but the rapid rate of vibration makes it very effective in driving through hard materials.

Grinding and cutting operations by ultrasonics on hard materials are similar to drilling. A coarse abrasive can be used at the start, then a finer grain is used until the correct finish is obtained. A good finish can be obtained at a lower temperature, thereby reducing damage. Ultrasonics can be used to cut glass, quartz, sapphire, germanium, silicon, diamonds, ceramics, and other hard materials.

Ultrasonic welding and soldering differs considerably from the regular process. No external heat is applied, only ultrasonic vibrations. Similar and dissimilar metals can be joined effectively. The amount of heat created by the ultrasonic vibrations is not sufficient to melt the metals being joined; therefore, the bond occurs by other means.

REVIEW QUESTIONS

1. What is the ultrasonic process?
2. What are the advantages in ultrasonic welding?
3. How is cutting an grinding accomplished in the ultrasonic process?

CHAPTER 24

Metal Spinning

The art of metal spinning has been practiced for many centuries and is still practiced both as a hobby and as a commercial method of shaping sheets of metal into circular shapes. In the spinning operation, a flat disk of metal is rotated at high speed, and pressed and shaped by means of hand tools against a revolving form mounted on a lathe. Bowls, plates, cooking utensils, and similar shapes can be shaped in this manner. Metal spinning is preferred to drawing between dies for the manufacture of a single article in limited numbers. Dies are expensive, and forms of spinning are cheaper and more easily made; the forms of spinning are seldom damaged. Dies are preferred for production work in which articles are produced in large quantities.

MATERIALS

Iron, soft steel, pewter, copper, brass, zinc, aluminum, silver, and britannia are all suitable for spinning. The metal-spinning operation requires a certain amount of physical work and skill. The various methods of spinning require different techniques, and operators specializing in a given metal may not be able to spin other metals satisfactorily. Faulty spinning is evident when the metal on the finished product is stretched to an uneven thickness. Good spinning technique results in a finished piece with variations in thickness of less than 25 percent of the original thickness. Of course, the percentage may vary with the character of the form, as some pieces are more difficult to spin. Copper and brass require more pressure than most other metals, such as britannia, because they are harder and offer more resistance to the forming tool.

The thickness of the metal and the kind of metal determine the rotation speed. Considerable variations in rotation speeds are required for the varying conditions and the different metals. The recommended speeds for the following metals are:

1. Iron and soft steel (300 to 600 r/min.).
    a. For $\frac{1}{16}$-inch stock (400 r/min.).
    b. For $\frac{1}{32}$-inch stock (600 r/min.).
2. Silver and britannia metal (800 to 1000 r/min.).
3. Copper (800 to 1000 r/min.).
4. Brass and aluminum (800 to 1200 r/min.).
5. Zinc (1000 to 1400 r/min.).

EQUIPMENT

Metal spinning requires slightly more rugged equipment than that required for wood turning. The common wood-turning lathe is not rugged enough to withstand the end thrust of the metal-spinning operations.

Lathe

A metal lathe or a sturdy wood-turning lathe is required for metal spinning. The lathe must have end-thrust bearings in the headstock. It must be of substantial construction, but back gears or a screw drive for the tool rest is not necessary (Fig. 24-1). A

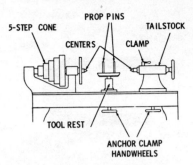

Fig. 24-1. The essential parts of a metal-spinning lathe.

five-step cone pulley is advantageous for a variety of speeds useful in spinning.

The *tool rest* should be movable on the lathe bed and held in any desired position by means of a handwheel clamp. The tool rest should be 12 to 15 inches in length, similar to the wood-turner's tool rest, with a series of closely spaced holes through the face of it. The small holes in the tool rest are necessary to receive the prop pin against which the hand tools are held while spinning. The prop pin should be 3 to 6 inches long, and made of ¾-inch-diameter steel turned down on one end to fit loosely in the holes in the tool rest.

The *tailstock center* for a metal-spinning lathe can be similar to that on a common lathe, but preferably it should be a revolving center (Fig. 24-2). The center should be ¾ inch in diameter and

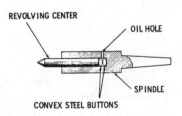

Fig. 24-2. A revolving tailstock center for a metal-spinning lathe.

about 6 inches long; it turns in a socket fitted to a taper hole in the tailstock. Note that the center is equipped with a double roller bearing, which considerably reduces friction.

Another type of bearing is shown in Fig. 24-3. The rollers revolve in opposite directions. The rollers do not rotate as fast as ball bearings, because they receive their motion nearer the center of rotation. Roller bearings are preferred to ball bearings—the ball bearings wear rapidly because of the tremendous amount of end

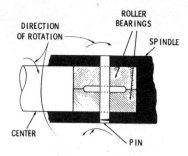

Fig. 24-3. Revolving tailstock center with a roller-thrust bearing.

thrust that results from the great amount of pressure necessary to force the center against the metal to prevent slipping. A center with a threaded end can be used, rather than a pointed center (Fig. 24-4). Back centers of various lengths and shapes can be turned onto the threaded end. The flats on the collar are provided for a wrench to be applied, to prevent rotation while the back center is being turned on.

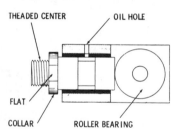

Fig. 24-4. Threaded revolving lathe tailstock center with a roller-thrust bearing.

Spinning Tools

A large variety of spinning tools are required in metal spinning. The spinning tools are made of tool steel forged to the required shapes from round steel, ranging from ½ inch to 1½ inches in diameter. Tools of aluminum bronze are generally used for spinning iron and steel, zinc, and aluminum. An alloy of copper (90 percent), aluminum (85 percent), and iron (2 percent) permits faster work than can be done with steel tools. Spinning tools are about 3 feet in length from the tip of the tool to the end of the handle (Fig. 24-5). The length is necessary to provide the large amount of leverage required to exert the great amount of pressure on the work for spinning. The wooden handle of the tools is usually

454

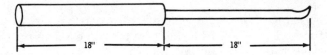

Fig. 24-5. Illustrating proper length of a metal-spinning hand tool.

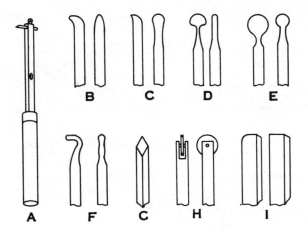

Fig. 24-6. Metal-spinning hand tools: (A) swivel cutter; (B) ball and Point;
(C) smoothing; (D) fish tail; (E) ball; (F) hook; (G) diamond point;
(H) beading; (I) skimmer.

about 2 inches in diameter and 18 inches in length. The long
wooden handle is necessary because the handle is held under the
armpit of the operator, so that sufficient physical power can be
applied to exert adequate pressure for spinning the metal.

The various spinning tools are shown in Fig. 24-6. The ball-and-
point tool is used most frequently. It is used to start the work and to
bring it to the approximate shape of the form. The ball-and-point
tool can be used for several operations because many different
shapes can be formed by merely turning the tool. Any rough
surfaces caused by the ball-and-point tool can be smoothed with
the smoothing tool. The fish-tail tool is used to flare the end of the
shell from the inside. A back stick made from a piece of hardwood
about 24 inches in length is another useful spinning tool.

455

Chucks or Forms

The shape of the chuck should conform to the desired shape of the finished article. This is the form that determines the shape of the shell.

A *solid chuck* is a plain form used for simple metal spinning (Fig. 24-7). Solid chucks are used when the shape of the work is such that it can be withdrawn from the chuck.

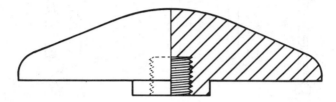

Fig. 24-7. A solid plain chuck, or form, used for simple metal-spinning work.

A *sectional chuck* is used when the shape of the work is such that it cannot be withdrawn from the chuck; the work is separated from the chuck by removing one section of the chuck at a time after the work is completed (Fig. 24-8).

A *plug* is a type of sectional chuck that is used where the shell has projections or shoulders at both ends and no bottom—for example, an hour glass (Fig. 24-9). It is made in two halves. One half is turned to take the shell from one end to the center of the smallest

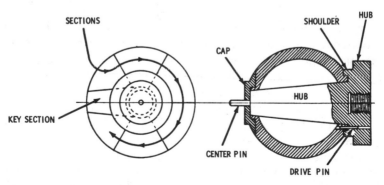

Fig. 24-8. A sectional chuck, or form, used for metal-spinning work that requires the chuck to be removed in sections after spinning is completed.

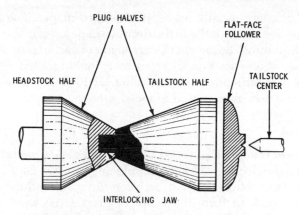

Fig. 24-9. A plug chuck, or form, and a flat-face follower.

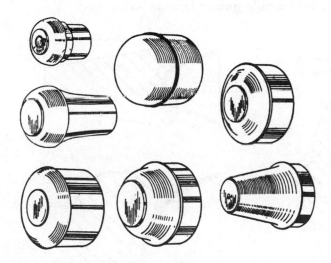

Fig. 24-10. Various shapes of chucks, or forms, used for metal spinning.

diameter. A hole is bored in the end of this section. The end of the second half is then fitted to the first half, and turning of the second half is completed. The plug is removed, after spinning, by cutting out the bottom of the shell, permitting the tail half of the plug to be removed.

457

The chucks are usually made of kiln-dried maple. The maple block is turned on the lathe to the desired shape to form the chuck for metal spinning. Some chucks are made of cast iron or steel. A chuck may be encased or armored by spinning a shell over the form and cementing it to the form. Fig. 24-10 illustrates some of the typical wood chucks used in metal spinning.

Followers

A follower is a block of wood used to hold a sheet-metal blank against the chuck or form for metal spinning (see Fig. 24-9). When the lathe is in motion, friction between the work and the chuck causes the work to turn; the follower presses the work firmly against the chuck. The hole in the center of the follower is to engage the revolving center. The follower should be made with the largest possible surface bearing on the work.

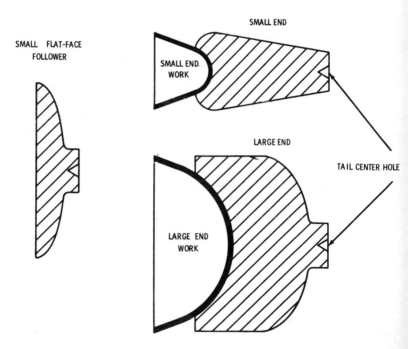

Fig. 24-11. Various followers used in metal spinning and their application.

Hollow followers are used on shells that do not have flat bottoms. The hollow part of the follower is shaped to fit the ends of the chucks with which they are to be used (Fig. 24-11). When a hollow follower is used in spinning, a small flat follower is used to hold the blank against the end of the chuck until the spinning has progressed enough to permit the stock to be engaged by a hollow follower (see Fig. 24-11).

HOW TO SPIN METAL

The design for a cover (Fig. 24-12A) can be used to illustrate metal spinning. In this instance, more than one chuck and follower are required to bring the circular disk blank to the desired shape. The correct size of blank required for the design can be formed by trial and error. The required chucks and followers are shown in Fig. 24-12B.

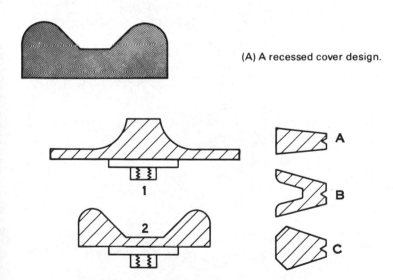

(A) A recessed cover design.

(B) Chucks and followers required to produce recessed cover.

Fig. 24-12. A typical metal-spinning design and the chucks and followers required to produce it.

Mounting the Work

Chuck No. 1 (see Fig. 24-12B) should be turned onto the lathe spindle. The blank should be pressed by hand, against the chuck, and the tailstock center turned against the center of the follower with just enough pressure to hold the blank in place. The blank is centered by pressing a hardwood stick against its edge after starting the lathe. As soon as the blank is centered, the pressured of the center against the blank should be increased gradually, and clamped in place. *Caution:* "Hopping in" a blank, or placing a blank in the lathe while the lathe is in motion, is a dangerous practice that should not be followed. The blank may become a dangerous missile, as it is likely to fly across the room, moving at a high rate of speed and turning at the same time.

The surface of the work should be lubricated with soap or beeswax before starting to spin. Lard oil with white lead is a less expensive substitute for beeswax. Mutton or beef tallow can be

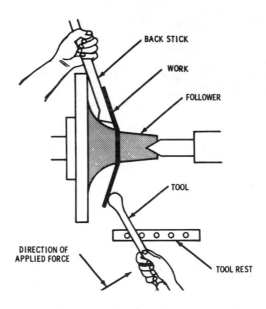

Fig. 24-13. The proper method of holding the metal-spinning tool. The handle of the tool should be placed under the right armpit, and the back stick should be held in the left hand.

used on most metals, and petroleum jelly mixed with graphite is another lubricant. The revolving blank should be rubbed lightly with the lubricant.

The Spinning Operation

The prop pin on the tool rest should be adjusted to a point near the blank so that good leverage for the spinning tool can be obtained. The operator should hold the handle of the tool under the right armpit, using the spinning tool as a lever and the prop pin on the tool rest as a fulcrum (Fig. 24-13). As pressure is applied, the metal disk is forced slowly against the chuck.

The spinning tool should be moved in and out constantly and should not be permitted to rest in one spot. Pressure should be applied only when moving toward the edge of the blank; pressure should not be applied when moving toward the center. The back stick, held in the left hand, should be pressed firmly against the reverse side of the metal at a constantly changing point opposite the tool (see Fig. 24-13). The back stick prevents wrinkling of the work as it is worked toward the edge of the disk.

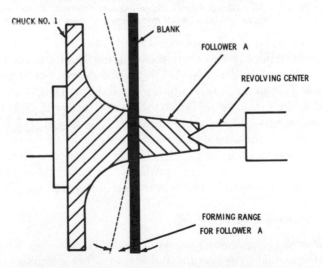

Fig. 24-14. After properly mounting the metal blank for the recessed cover between chuck No. 1 and follower A, it should be worked toward the chuck only as far as the dotted line before substituting follower B. See Fig. 24-12.

461

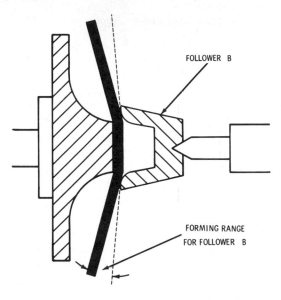

FOLLOWER B

FORMING RANGE
FOR FOLLOWER B

Fig. 24-15. After substituting follower B and follower A, spinning is continued until the work has the same form as chuck No. 1.

After the work is mounted between chuck No. 1 and follower *A*, the metal should be worked back a short distance before follower *B* is substituted (Fig. 24-14). Follower *B* gives a better grip on the work. Spinning can be continued until the work is in the shape of chuck No. 1 (Fig. 24-15). Then the work is reversed, and chuck No. 2 is substituted, along with follower *C* (Fig. 24-16). The final stage of spinning consists of shaping the outer part of the work to the corresponding part of the chuck (Fig. 24-17). Sometimes a compound lever arrangement is utilized so that additional tool pressure can be exerted by the operator on the workpiece (Fig. 24-18).

Annealing and Pickling

During the spinning process, a gradual hardening, which renders the metal brittle and unsuitable for further spinning, can occur. When the metal reaches this state, the operator should remove the work from the lathe and anneal it before continuing the spinning process.

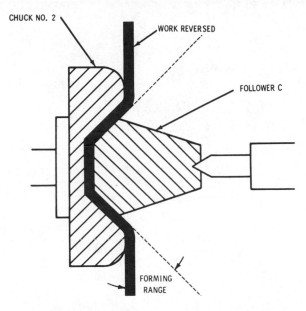

Fig. 24-16. After the shape of chuck No. 1 has been achieved, chuck No. 2 and follower C are substituted. The dotted line indicates the position of the work at the beginning of this operation.

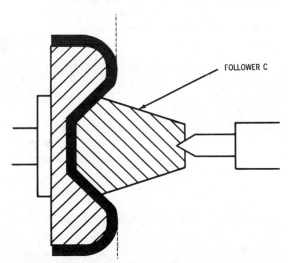

Fig. 24-17. The final shape of spinning consists of shaping the outer part of the recessed cover to the chuck.

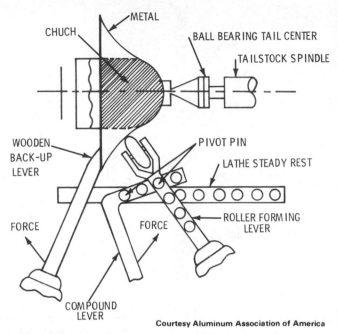

METAL

CHUCH

BALL BEARING TAIL CENTER

TAILSTOCK SPINDLE

WOODEN
BACK-UP
LEVER

PIVOT PIN

LATHE STEADY REST

FORCE

FORCE

ROLLER FORMING
LEVER

COMPOUND
LEVER

Courtesy Aluminum Association of America

Fig. 24-18. A manual power-lathe spinning setup. With this type of compound lever arrangement the operator can exert additional tool pressure on the workpiece.

The operator can determine when the metal should be annealed by the variation in effort required to hold the spinning tool under his armpit. The metal should be fluted or dented by hammering with a rawhide hammer to relieve the stresses and to prevent cracking during the annealing process.

After it has been annealed, the oxide scale formed in the annealing process should be removed with a pickling solution. Then the work should be washed thoroughly in running water and permitted to dry.

SUMMARY

Dies are preferred for production work in which articles are produced in large quantities, but for a limited number of a single article, most manufacturers prefer metal spinning. Dies are expensive while the forms of spinning are cheaper and more easily made.

Iron, soft steel, copper, brass, zinc, and aluminum are just a few metals suitable for spinning. The various methods of spinning require different techniques, and operators specializing in a given metal may not be able to spin other metals satisfactorily. Copper and brass require more pressure than most other metals, such as britannia, because they are harder and offer more resistance to the forming tool.

Metal spinning requires more rugged equipment than that required for wood turning. The common wood-turning lathe is not rugged enough to withstand the end thrust of the metal spinning operation. The lathe must have end-thrust bearings in the headstock.

REVIEW QUESTIONS

1. What metals are best suited for spinning?
2. What is the difference between a metal-spinning lathe and a wood-turning lathe?
3. What metals are generally used to make spinning tools?
4. What is a follower?

CHAPTER 25

Automated
Machine Tools

Automation is sometimes confused with the use of computers in industry. The term *automation* is actually closer in meaning to the control of machine tools, which is only a single use for computers in industry. Computers are an aid in a wide variety of industries in the areas of inventory control, process control, and numerical control of machine tools.

The computer is used in effective inventory-control programs to provide management with the facts on which decisions can be based. It is used to forecast high-demand items, to adjust production to market, and to parallel more closely other trends with production schedules.

The computer is used in process control in industry or manufacturing to monitor by means of sensors, which measure whatever is going on. These measures are fed into a computer for control of the steps in the process. Signals from mechanical and electrical instructions, which are used to point out certain conditions and timing in the process or product, are used to control every industrial process. Normally, the operating personnel interpret these signals and transform them into control actions, but the input elements of the computer are designed to accept these same control signals so that the computer can interpret them and translate them into effective control actions. For example, industrial transducers that change pressure into voltage are used to convert these input signals into suitable voltages. The selected input groups are channeled to the computer logic in the correct sequence.

Control outputs from the computer can be converted to the required control actions, such as setting switches or turning valves. Thus, the computer actually controls the process, its timing, and its operation.

Numerical control for a system of automation uses a punched tape in which the holes represent coded instructions for the machine tool (Fig. 25-1). These coded instructions tell the machine

Courtesy Moog Hydro-Point

Fig. 25-1. Programming and preparing tape is simple and time saving. It involves deciding on the number of workpieces per load and their location on the subplate; preparing the tool and program sheets from the piece part drawing with coordinate dimensions; and punching tape using conventional tape preparation equipment.

exactly where to drill a hole, and they give the exact size of the hole. This results in an automatic operation of the machine tool (Fig. 25-2). This chapter is concerned chiefly with numerical control.

BASIC PRINCIPLES OF NUMERICAL CONTROL

One important aspect of computers and computer-like techniques is the numerical or digital control of machine tools. Numerical control is concerned with how the machine director or controller responds to commands expressed by numbers. The unit causes the controls of the machine, such as handwheels or cranks, to be moved by means of power, rather than moved manually by an operator (Fig. 25-3).

As automation is a method that is used to avoid physical labor, it is usually accomplished by designing machines that repetitively perform either single or multiple operations that would normally be performed by a machine operator. Numerical control is a

Courtesy Moog Hydro-Point

Fig. 25-2. A row of numerically controlled machining centers with 24-station random selection automatic tool changers.

469

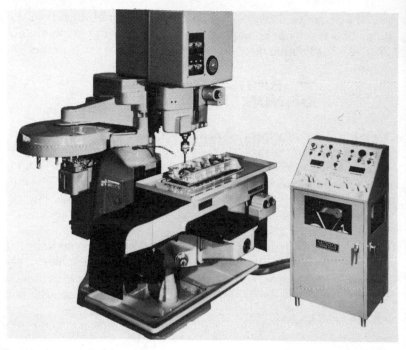

Courtesy Moog Hydro-Point

Fig. 25-3. This numerically controlled machining center is comprised of a milling machine and control unit. The control unit causes the controls of the milling machine to move by means of power rather than manually by an operator.

method of industrial automation where the method used to avoid physical labor, instead of being a "fixed" repetitive method, can be changed at will by means of variable programming.

Numerical control is a type of programmed automation. Consider the common drill press, for example, as a power-driven tool. The vertical spindle of the drill press moves up and down in relation to the worktable, which normally is fixed in position. The workpiece to be drilled can be moved into position by maneuvering it on the worktable until the spot where the hole is to be drilled is directly below the twist drill.

Consider a 500-tooth gear and a lead screw pitch of one inch per revolution, so that the table moves one inch when the gear and shaft are rotated one revolution. If the gear is moved only 10 teeth, the slide will move only 0.02 inch; for a movement of one tooth, the slide will move 0.002 inch.

Stops are used on the gears; if there are 10 stops spaced equally around a gear with 1000 teeth, an amount of movement of the worktable (in decimal parts of an inch) can be selected. The first stop will let the gear rotate $\frac{1}{10}$ revolution, or 100 teeth. If a train of four gears is involved, each having 10 stops, and the gears are connected by a gear reduction of 10 to 1, the first gear and the interposers will control the table movement in one-inch steps within a 10-inch measure. The second gear will control the table movement in 10 steps of 0.001 inch. It is possible, by selecting the proper stop on each of the four gears, to obtain stops that control movement to the precision of a four-digit number, such as 2.357 inches.

Switches can be set up manually; or a punched tape and a tape reader can be used to decode a set of holes for each digit. Thus, the slide table on the drill press will move the correct distance for the dimension as was programmed in the tape.

Several other systems can be used, two of which use electronics. One system uses voltages that are related to digits in the dimensions. The worktable has a transformer that produces a voltage proportional to its movement; the movement of the worktable stops when the two voltages are equal.

Another electronic system uses pulses; the dimension is given in terms of the number of electrical pulses. For example, if the dimension is 42.3715 inches, there are 423,715 pulses to be counted. In this system the drive motor stops when the number of pulses counted equals the number set up in the predetermined counter.

Pulses for counting can be produced by several methods of generating the pulses. One of the most accurate methods is the use of a finely graduated scale on a bar of glass connected to the worktable. In this system, these graduations are a series of fine parallel lines on the glass bar. Using a light source on one side and a photocell on the other, a pulse is produced for each line on the glass bar when the beam of light to the photocell is broken by the line. Line width and spacing between lines are equal; as many as 10,000 lines per inch are possible.

A different method uses a steel bar with magnetic spots, on the worktable; a magnet pickup coil is mounted near the magnetic spots on the bar. When these spots pass by the pickup coils, a voltage pulse is generated, and pulses are then counted by special circuits.

The pulse number is set up into a predetermined or preset counter. There is a difference between the pulse number and the number registered from the scale, but a closed control relay circuit supplies the power to an electric motor which, in turn, drives the gear and the lead screw, moving the table in the proper direction for the pulses to be counted. When the two pulse numbers are the same, this is recognized, and the motor stops at this point. As a result, the worktable stops in the proper position.

Many applications of numerically controlled manufacturing processes can be found, including reversing tape rolling mills for aluminum and steel, where the ingot moves back and forth through a series of rollers which force the metal into thin bars and then into thin sheets. As the opening between the rollers must be changed and made smaller and smaller in this process, close control of the gage screws is required; these are turned until the bar is finally reduced to the desired thickness. The rate and amount of turning down varies with the material, temperature, and other characteristics. This action is controlled by a program using either a punched tape or a punched card.

Numerically controlled ultraprecision inspection and measuring machines are in use. The machine tool industry has made wide use of this principle, and has found that it can be applied to many processes and operations including drilling machines, boring machines, milling machines, lathes, shapers, punch presses, flame cutters, and welders.

PREPARATION FOR NUMERICAL CONTROL

The Armour Institute of Chicago was one of the first to study the field of numerical control; they applied it to a lathe. The Massachusetts Institute of Technology applied numerical control to a milling machine in 1952. The operation used punched paper tape; tape-preparation equipment and a tape reader were used to place the numerical-control information into the machine tool director. Several types of equipment are now available as a result of expansion of the employment of numerical control in the industrial field.

The first step in numerical control planning is to start with the raw data or engineering drawing of the part. The process planner must study the part, determine just how it is to be made, and plan to use the standard tools, such as drills, reamers, cutters, and holding fixtures; or provide for special tooling if it is required. Special tools, which are usually holding fixtures, are sometimes required when the part cannot be mounted in the usual manner on the worktable of the machine tool by means of the fixture in stock. After recording all the necessary information on a planning sheet, the process planner must check the sheet for accuracy to make certain that the processing steps are in the correct sequence and will work.

The next step, after completion of the planning sheet, is to have it typed on a machine that produces a punched paper tape or "control tape" (Fig. 25-4). The control tape is checked carefully

Courtesy Moog Hydro-Point

Fig. 25-4. Conventional tape preparation unit. This unit is usually located away from the machining area.

473

for accuracy of interpretation from the source copy and for machine accuracy to make sure that it will work. The control tape can be retained as a "master" control tape and a duplicate tape made for use in the shop. The tape is placed in the tape reader on the machine director to run the machine for production of the required parts, (Fig. 25-5). Usually the tape is used to run the machine through all the steps in the numerical-control operation without any work on the worktable to make sure that a step has not been overlooked, or incorrectly programmed, and that production is possible.

To illustrate preparation for the numerical-control operation, the drawing of the casting (Fig. 25-6) can be used as the starting point for programming. All required dimensions of the part are taken from the drawing and placed on the process planning sheet,

Courtesy Moog Hydro-Point

Fig. 25-5. Control unit for a machining center. In addition to providing the capability of making the initial tape, this unit also enables the operator to make changes or corrections easily and quickly right at the machine, thus eliminating the need to send the original tape back tot he programming department for editing. This unique advantage can reduce setup time by as much as 40 percent.

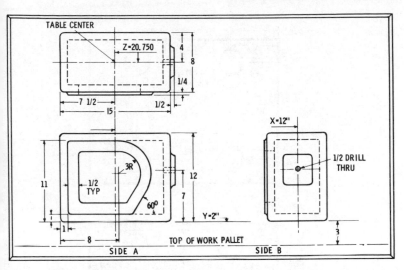

Fig. 25-6. Engineering drawing of a casting that is to be produced by numerical-control operations.

along with the other information. In addition to the X and Y coordinates (Fig. 25 7) a considerable amount of information is required. The tool group and number and the auxiliary functions to be performed are all required, as well as the X and Y coordinates for each step. Programming can be one-quadrant or four-quadrant, as shown in Fig. 25-7.

The final program or numerical-control tapes vary in length, depending on the number of operations, but most tapes range from 19 to 60 inches in length. Some tapes are much longer.

The Electric Industries Association on numerical controls has a standard one-inch tape with as many as eight holes for coding. Several other organizations, such as the National Machine Tool Builders Association and the Aerospace Industries Association have agreed to adopt these same standards for their organizations. Typical coding, in which the binary coded decimal system is used, is shown in Table 25-1.

The unpunched paper tape has an overall width of 1.000 inch plus or minus 0.003, a thickness of 0.004 inch plus or minus 0.003 inch, and can be used for recording six, seven, or eight levels of information across the tape.

475

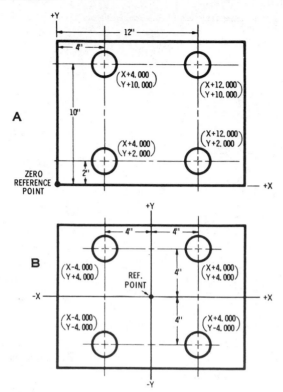

Fig. 25-7. Programming a part for production by numerical-control methods: (A) one-quadrant programming; (B) four-quadrant programming.

A standard format of the program in a punched tape permits the tape prepared for use on one machine to be used on other machines. A tape prepared for a given two-axis numerically controlled machine should work on all two-axis numerically controlled machines.

Various programming formats are possible. Fig. 25-8 compares four methods of programming that can be used for a numerical-control tape prepared for the same item. The four systems are: *Fixed Sequential*, *Block Address*, *Word Address*, and *Tab Sequential*. Also, a format is shown for a universal tape, which can be used for any one of the other systems.

Table 25-1. Binary Coded Decimal System for Numerical Control

Standard Track Numbers (8 7 6 5 4 . 3 2 1)	Digit or Letter Codes	Standard Track Numbers (8 7 6 5 4 . 3 2 1)	Digit or Letter Codes
6 .	0	7 5 . 3 2 1	p
. 1	1	7 5 4 .	q
. 2	2	7 4 . 1	r
5 . 2 1	3	6 5 . 2	s
. 3	4	6 . 2 1	t
5 . 3 1	5	6 5 . 3	u
5 . 3 2	6	6 . 3 1	v
. 3 2 1	7	6 . 3 2	w
4	8	6 5 . 3 2 1	x
5 4 . 1	9	6 5 4 .	y
7 6 . 1	a	6 4 . 1	z
7 6 . 2	b	7 6 4 . 2 1	. (period)
7 6 5 . 2 1	c	6 5 4 . 2 1	, (comma)
7 6 . 3	d	7 6 5 .	+ (plus)
7 6 5 . 3 1	e	7 .	− (minus)
7 6 5 . 3 2	f	6 5 . 1	√ (check)
7 6 . 3 2 1	g	7 6 5 4 . 3 2 1	Delete
7 6 4 .	h	8 .	End of Block
7 6 5 4 . 1	i	4 . 2 1	End of Record
7 5 . 1	j	5 .	Space
7 5 . 2	k	6 4 . 2	Back Space
7 . 2 1	l	6 5 4 . 3 2	Tab
7 5 . 3	m	7 6 5 4 . 3	Upper Case
7 . 3 1	n	7 6 5 4 . 2	Lower Case
7 . 3 2	o		

ELECTRONIC CONTROL OF MACHINE TOOLS

Automation means control of manufacturing. In electronic control of machine tools, a recording is made of the various movements required of the machine tool for a specific series of opera-

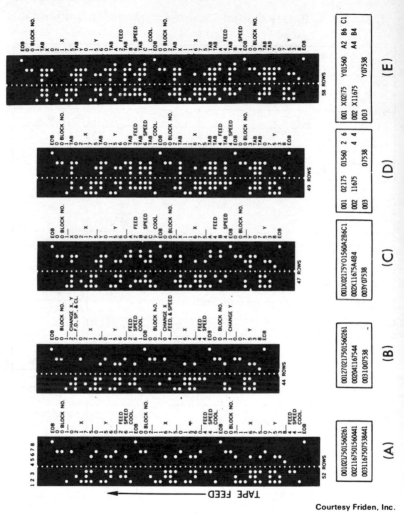

Courtesy Friden, Inc.

Fig. 25-8. Various types of punched paper-taper format: (A) fixed sequential, (B) block address, (C) word address, (D) tab sequential, and (E) universal.

tions that are required to produce a machined part. The recording, which is translated from a drawing of the part, is recorded by means of either a magnetic tape or a punched paper tape.

Typical modern units used for numerical control of machine

478

Courtesy Ex-Cell-O Corporation

Fig. 25-9. Automatic numerical-control system.

tools are shown in Fig. 25-9. Four steps involved in their use are: (1) development of the blueprint, (2) planning, (3) preparation of a control tape, and (4) running the tape on the machine.

The numerical-control systems are basically incremental point-to-point positioning systems of the on-off tape using binary-coded decimals. As many as five axes can be controlled either simultaneously or sequentially with the basic system. System resolution can be either 0.001 inch or 0.0001 inch, with only nominal differences in the electronics.

The block diagram shown in detail in Fig. 25-10 shows one axis of the control. All the control elements except the tape reader, distributor, and sequence control are repeated for the other axes. This is typical of a machine tool system that machines automatically and performs single-point boring operations from either end with infinite center locations and variable bore diameters, within the range of the machine. Tape-control positioning equipment is used with the capability of controlling the size of bores during the machining cycle. Table 25-2 is a summary of the numerical-control specifications that are possible in the system shown in Fig. 25-10.

Three traditional axes are used: longitudinal (X), horizontal (Y),

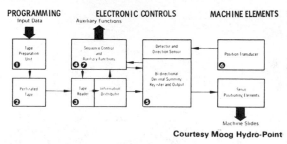

Fig. 25-10. Numerical-control system block diagram.

and vertical (Z). A fourth axis of control is necessary for boring holes of varying sizes. The fourth axis is the tool or radius axis (R) which permits control of the programming of an infinite number of boring diameters.

The addition of the tool or radius axis required the development of a tool control mechanism capable of accepting numerical commands. In this device, an electric servomotor is used to position the boring bar to change boring size in response to punched-tape signals.

There are three modes of operation:

1. *Dial In*, or *Manual Mode*, is used for very short runs. The decade counter dials are set to the required dimensions. The axis direction switches are set, and the "start" button pushed. The dimensions are displayed decimally. When the "manual-start" button is pressed, the table moves to the dimensions immediately.

2. *Semiautomatic*, or *Block Mode*, is used for tape verification and general checking. The tape is inserted in the tape reader and the "read" button pushed. The photoelectric reader reads the tape at 100 times per second, and the dimensions are displayed decimally on the decade counter tubes of the summing register, where they can be compared with the blueprint or planning sheet dimensions. In either manual or block mode, the spindle (if applicable) can then be actuated.

3. The *Automatic Mode* is normally used to operate the machine tool automatically through programmed functions. The tape is inserted in the reader and the "read" button pressed. The photoelectric reader reads one block or a com-

Table 25-2. Summary of Numerical-Control Characteristics

	GENERAL	TYPICAL MODELS*				
		M23	M33	M44	M54	
Number of Axes	Per Requirements	2	3	4	5	
Travel	Virtually unlimited in all axes.	99.999"	99.999"	X 99.9999" Y 99.9999" Z 99.9999" R 0.9999"	X 99.9999" Y 99.9999" Z 99.9999" R 0.9999"	
Resolution	0.0001 in. (all axes if desired).	0.001 in. all axes	0.001 in. all axes	0.0001 in. all axes	X 0.001 in. Y, Z, R 0.0001 in.	
Positioning Speed	Unlimited depending upon resolution only.	180 in./min.	180 in./min. Z-210 in./min.	100 in./min.	120 in./min.	
Tape	1-in.-wide, 8-channel Al-Mylar laminate.	Same	Same	Same	Same	
Tape Reading Speed	100 lines/sec.	Same	Same	Same	Same	
Auxiliary Functions	Per requirements	11	11	20	20	
Feed Rates	Large number available in any or all axes.	—	—	0.5 to 20 in./min.in 31 steps	0.5 to 10 in./min. in 16 steps	
Operation Modes	Automatic, semi-automatic, manual.	Same	Same	Same	Same	
Voltage Requirement	115AC, Single phase.	Same	Same	Same	Same	

*Rheem Numerical Positioning Controls can be applied to almost all machine positioning requirements. These typical models are only representative. Configurations for other applications are available.

Courtesy Ex-Cell-O Co.

plete machine command. At the end of the reading phase, the slide axes move to position simultaneously, and the spindle is then activated. The operator can always check the accuracy of the system by watching the position display.

The system block diagram consists of three parts; programming, electronic controls, and the elements of the machine tool itself. Any number of axes can be controlled simultaneously or in a specified sequence. Each axis of control is complete in a chassis that slides into the main cabinet.

The output from this system can control electric drive motors or hydraulic servo valves that control hydraulic motors or pistons. Feed-rate control can be added for applications that require various controlled feed rates.

Tape Preparation

In reference to Fig. 25-10, a tape preparation unit (1) is used to prepare a punched paper tape (refer to Fig. 25-4) (2). The input unit is a tape reader (3) that operates at a speed of 100 lines per second. No machine time is lost for data input. Tape information is distributed to auxiliary functions and sequence control (4) and to the summing register's computer (5) by electronic switching.

The data storage medium in this instance (Fig. 25-11) is standard one-inch, eight-channel paper, *Mylar*, or *Al-Mylar* tape. The tape code is EIA approved binary-coded decimal, and the tape format is tab sequential. The tab markers (code holes in the tracks number 2, 3, 4, 5, and 6) are punched to separate information code words, such as X-axis or Y-axis.

The end-of-block code can be punched at any line of the tape denoting that no more information is forthcoming for that machine command. Only the tab marker need be punched when no movement is required of an axis. It is unnecessary to add unused axis dimensions. Thus in the operation shown in block No. 2, only the tab markers are punched for X and Y.

Control

The tape reader reads one complete operation (distance to be moved and direction of motion for each axis) into the bidirectional

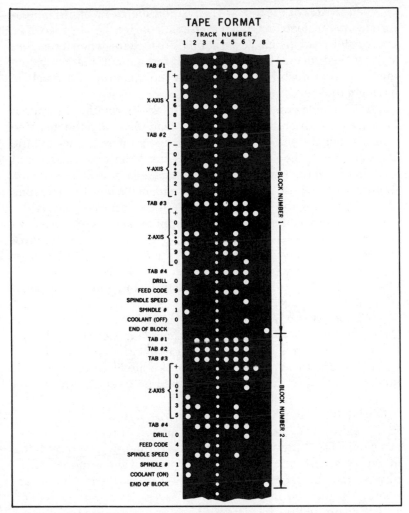

Fig. 25-11. The tab-sequential tape format used with the machine.

register (5) whose output controls drive the axes simultaneously in the command direction at rapid-traverse speed. The position transducers (6) convert motion into numbers for feedback signals indicating direction and distance of motion. The register also counts the distance moved, compares it to what it should be, and makes corrections.

483

When all axes of the tool are in position, an "In-Position" lamp is turned on. An electric signal is then available for the tool feed. The control will read the next operation from the tape when it receives a "tool-feed cycle finished" signal for an automatic cycling machine, or a push-button signal for the operator of a machine having a tool-feed cycle.

The sequence control (4 and 7 in Fig. 25-10) divides the system into two basic operating phases—the read phase and the machine phase. When the end-of-block signal appears after a block of tape has been read, the sequence control stops the tape reader and switches from the read phase to the machine phase. The machine phase "start" signal allows axis movement. When the machine element reaches the programmed position, a move relay is de-energized, and a coincidence signal appears at the sequence control. When all axes have reached coincidence, the sequence control switches from machine phase back to read phase, and the cycle is repeated for every block programmed on the tape. An end-of-tape signal prevents the sequence control from entering the read phase again.

At the beginning of the read phase, a tape-start signal is sent to the summing register to clear any spurious counts and to inhibit the borrow and carry function, because decimal information is entered into each decade in parallel. At the end of the read phase, a tape-stop signal is transmitted to the summing register to free the counter for the machine phase.

During the read phase, each of the several decades of the summing register receives decimal information in parallel from the electronic distributor. This programmed dimension represents the incremental distance to the next slide position and is momentarily displayed in the decimal counting tubes prior to the machine-start phase. During the machine phase, pulses are received by the register from the position transducer. The pulses are serially fed into the least significant decade of the bidirectional counter; these pulses return the counter to zero.

Transducers

The sensing elements of the position transducer are two magnetic heads that sense the rotation of the shaft. The unit can be applied either to precision ball-loaded lead screws or to precision measuring racks with pinions.

During the machine phase, the transducers feed back digital pulses, indicating distance and direction of movement. These pulses are fed serially into the least-significant decade of the bidirectional counter. The counter counts down until the slide is at a predetermined distance from the stopping point. At this time the slide velocity is precisely set to the controlled approach rate. At the stopping point, the last pulse to the counters stops the slide drive motors, activates the slide lock, and triggers the next function if applicable.

The number of auxiliary functions that can be controlled is almost unlimited. Examples of these functions are spindle or tool

Courtesy Moog Hydro-Point

Fig. 25-12. Peck-drilling. A pecking action is used to facilitate deep hole drilling. The cutting time between pecks is preset on a timer, eliminating the necessity to program each peck depth. At each peck the tool is withdrawn (at 300 IPM) to clear the chips and returns (at 300 IPM) to 0.015 inch above the previous cut. This assures maximum tool life and gives accurate deep holes, as small as 0.014 inch in diameter, without tool breakage.

Fig. 25-13. Pocket milling can be carried out with guaranteed accuracy in all three axes. The backlash-free positioning system permits milling in both directions on each axis; and climb milling can be programmed to produce the best possible surface finish.

Fig. 25-14. Boring. A super-precision quill fit coupled with guaranteed spindle runout less than 0.0002 inch TIR and positive tool orientation assures close tolerance bored holes. Hole size and geometry can be held within 0.0002 inch under production conditions. This "canned Cycle" allows tool withdrawal at programmed feed rate, thus eliminating taper and tool marks.

Courtesy Moog Hydro-Point

Fig. 25-15. Fly-cutting. Excellent finishes can be obtained using up to 5-inch diameter fly-cutters. Up to 3-inch diameter can be loaded in adjacent stations.

Courtesy Moog Hydro-Point

Fig. 25-16. Heavy milling cut. Machines will remove 2.3 cubic inches per minute of mild steel.

Fig. 25-17. Coordinated tapping provides a fast tapping cycle, where spindle speeds and feeds are coordinated under tape-control, to match the pitch of the thread. A compensating tension/compression tap holder is used for the operation.

Fig. 25-18. Drill cycle. The standard drill cycle can be used to perform boring and reaming operations where rapid retraction of the tool is acceptable.

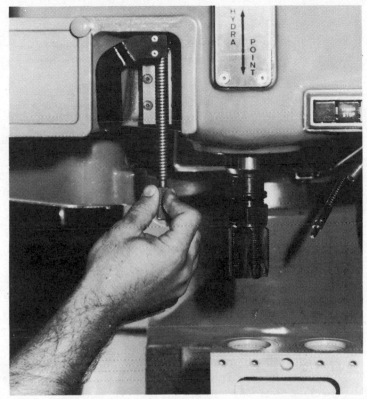

Courtesy Moog Hydro-Point

Fig. 25-19. The lead-screw tapping operation will produce class-3 threads with very accurate depth control. A precision-ground lead screw is hydraulically traced, and controls and feed rate of the tap in relation to the pitch and spindle speed. The only load applied to the tap is the torque required for cutting.

selection, spindle speeds, coolant on-off, and feed-rate control. Feed rate is generated as programmed from the tape for the desired feed-rate characteristics in applications where it is necessary. Machine sequencing and output controls are mechanized with relays to meet the requirements of a particular machine and its positioning elements.

Illustrated are a number of close-up views of machining operations (Figs. 25-12 thru 25-19) performed on the numerically controlled machining center illustrated in Figs. 25-2 and 25-3.

SUMMARY

Automation is a method that is used to avoid physical labor and is accomplished by designing machines that repetitively perform either single or multiple operations that would normally be performed by a machine operator. Numerical control, instead of a "fixed" repetitive method, can be changed at will by means of variable programming.

Various automation systems can be used. Switches, or punched tape and a tape reader, can be used to decode a set of holes for each digit. Other electronic systems utilize voltage pulses. Pulses for counting can be produced by several methods. One of the most accurate methods is the use of a finely graduated scale on a bar of glass connected to the worktable.

REVIEW QUESTIONS

1. What is automation?
2. Name the various automation systems used on machine tools.
3. Explain the preparation of tape.
4. What is the sensing element or transducer? What does it do in automation?

Appendix

Colors and Approximate Temperatures
for Carbon Steel

Black Red	990 °F	532 °C
Dark Blood Red	1050	566
Dark Cherry Red	1175	635
Medium Cherry Red	1250	677
Full Cherry Red	1375	746
Light Cherry, Scaling	1550	843
Salmon, Free Scaling	1650	899
Light Salmon	1725	946
Yellow	1825	996
Light Yellow	1975	1080
White	2220	1216

Nominal Dimensions of Hex Bolts and Hex Cap Screws

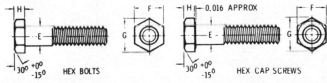

HEX BOLTS		HEX CAP SCREWS	
Nominal Size E	Width Across Flats F	Width Across Corners G	Head Height H
$1/4$	$7/16$	$1/2$	$11/64$
$5/16$	$1/2$	$9/16$	$7/32$
$3/8$	$9/16$	$21/32$	$1/4$
$7/16$	$5/8$	$47/64$	$19/64$
$1/2$	$3/4$	$55/64$	$11/32$
$5/8$	$15/16$	$1\,3/32$	$27/64$
$3/4$	$1\,1/8$	$1\,19/64$	$1/2$
$7/8$	$1\,5/16$	$1\,33/64$	$37/64$
1	$1\,1/2$	$1\,47/64$	$43/64$
$1\,1/8$	$1\,11/16$	$1\,61/64$	$3/4$
$1\,1/4$	$1\,7/8$	$2\,11/64$	$27/32$
$1\,3/8$	$2\,1/16$	$2\,3/8$	$29/32$
$1\,1/2$	$2\,1/4$	$2\,19/32$	1
$1\,3/4$	$2\,5/8$	$3\,1/32$	$1\,5/32$
2	3	$3\,15/32$	$1\,11/32$

Nominal Dimensions of Heavy Hex Bolts and Heavy Hex Cap Screws

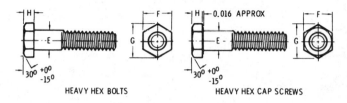

HEAVY HEX BOLTS   HEAVY HEX CAP SCREWS

HEAVY HEX BOLTS			HEAVY HEX CAP SCREWS	
Nominal Size	Width		Height	
	Across Flats F	Across Corner G	Bolts H	Screws H
$1/2$	$7/8$	1	$11/32$	$5/16$
$5/8$	$1\,1/16$	$1\,15/64$	$27/64$	$25/64$
$3/4$	$1\,1/4$	$1\,7/16$	$1/2$	$15/32$
$7/8$	$1\,7/16$	$1\,21/32$	$37/64$	$35/64$
1	$1\,5/8$	$1\,7/8$	$43/64$	$39/64$
$1\,1/8$	$1\,13/16$	$2\,3/32$	$3/4$	$11/16$
$1\,1/4$	2	$2\,5/16$	$27/32$	$25/32$
$1\,3/8$	$2\,3/16$	$2\,17/32$	$29/32$	$27/32$
$1\,1/2$	$2\,3/8$	$2\,3/4$	1	$15/16$
$1\,3/4$	$2\,3/4$	$3\,11/64$	$1\,5/32$	$1\,3/32$
2	$3\,1/8$	$3\,39/64$	$1\,11/32$	$1\,7/32$

492

Nominal Dimensions of Heavy Hex Structural Bolts

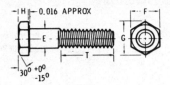

Nominal Size E	Width Across Flats F	Width Across Corners G	Head Height H	Thread Length T
$1/2$	$7/8$	1	$5/16$	1
$5/8$	$1\,1/16$	$1\,15/64$	$25/64$	$1\,1/4$
$3/4$	$1\,1/4$	$1\,7/16$	$15/32$	$1\,3/8$
$7/8$	$1\,7/16$	$1\,21/32$	$35/64$	$1\,1/2$
1	$1\,5/8$	$1\,7/8$	$39/64$	$1\,3/4$
$1\,1/8$	$1\,13/16$	$2\,3/32$	$11/16$	2
$1\,1/4$	2	$2\,5/16$	$25/32$	2
$1\,3/8$	$2\,3/16$	$2\,17/32$	$27/32$	$2\,1/4$
$1\,1/2$	$2\,3/8$	$2\,3/4$	$5/16$	$2\,1/4$

Nominal Dimensions of Hex Nuts, Hex Thick Nuts, and Hex Jam Nuts

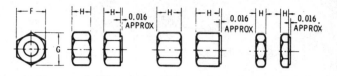

Nominal Size	Width Across Flats F	Width Across Corners G	Thickness		
			Hex Nuts H	Thick Nuts H	Jam Nuts H
$1/4$	$7/16$	$1/2$	$7/32$	$9/32$	$5/32$
$5/16$	$1/2$	$9/16$	$17/64$	$21/64$	$3/16$
$3/8$	$9/16$	$21/32$	$21/64$	$13/32$	$7/32$
$7/16$	$11/16$	$51/64$	$3/8$	$29/64$	$1/4$
$1/2$	$3/4$	$55/64$	$7/16$	$9/16$	$5/16$
$9/16$	$7/8$	1	$31/64$	$39/64$	$5/16$
$5/8$	$15/16$	$1\,3/32$	$35/64$	$23/32$	$3/8$
$3/4$	$1\,1/8$	$1\,19/64$	$41/64$	$13/16$	$27/64$
$7/8$	$1\,5/16$	$1\,33/64$	$3/4$	$29/32$	$31/64$
1	$1\,1/2$	$1\,47/64$	$55/64$	1	$35/64$
$1\,1/8$	$1\,11/16$	$1\,61/64$	$31/32$	$1\,5/32$	$39/64$
$1\,1/4$	$1\,7/8$	$2\,11/64$	$1\,1/16$	$1\,1/4$	$23/32$
$1\,3/8$	$2\,1/16$	$2\,3/8$	$1\,11/64$	$1\,3/8$	$25/32$
$1\,1/2$	$2\,1/4$	$2\,19/32$	$1\,9/32$	$1\,1/2$	$27/32$

Nominal Dimensions of Square Head Bolts

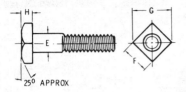

Nominal Size E	Width Across Flats F	Width Across Corners G	Head Height H
$1/4$	$3/8$	$17/32$	$11/64$
$5/16$	$1/2$	$45/64$	$13/64$
$3/8$	$9/16$	$51/64$	$1/4$
$7/16$	$5/8$	$57/64$	$19/64$
$1/2$	$3/4$	$1\,1/16$	$21/64$
$5/8$	$15/16$	$1\,21/64$	$27/64$
$3/4$	$1\,1/8$	$1\,19/32$	$1/2$
$7/8$	$1\,5/16$	$1\,55/64$	$19/32$
1	$1\,1/2$	$2\,1/8$	$21/32$
$1\,1/8$	$1\,11/16$	$2\,25/64$	$3/4$
$1\,1/4$	$1\,7/8$	$2\,21/32$	$27/32$
$1\,3/8$	$2\,1/16$	$2\,59/64$	$29/32$
$1\,1/2$	$2\,1/4$	$3\,3/16$	1

Nominal Dimensions of Heavy Hex Nuts
and Heavy Hex Jam Nuts

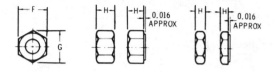

Nominal Size	Width Across Flats F	Width Across Corners G	Thickness Hex Nuts H	Thickness Hex Jam Nuts H
$1/4$	$1/2$	$37/64$	$15/64$	$11/64$
$5/16$	$9/16$	$21/32$	$19/64$	$13/64$
$3/8$	$11/16$	$51/64$	$23/64$	$15/64$
$7/16$	$3/4$	$55/64$	$37/64$	$17/64$
$1/2$	$7/8$	$1\,1/64$	$31/64$	$19/64$
$9/16$	$15/16$	$1\,5/64$	$35/64$	$21/64$
$5/8$	$1\,1/16$	$1\,7/32$	$39/64$	$23/64$
$3/4$	$1\,1/4$	$1\,7/16$	$47/64$	$27/64$
$7/8$	$1\,7/16$	$1\,21/32$	$55/64$	$31/64$
1	$1\,5/8$	$1\,7/8$	$63/64$	$35/64$
$1\,1/8$	$1\,13/16$	$2\,3/32$	$1\,7/64$	$39/64$
$1\,1/4$	2	$2\,5/16$	$1\,7/32$	$23/32$
$1\,3/8$	$2\,3/16$	$2\,17/32$	$1\,11/32$	$25/32$
$1\,1/2$	$2\,3/8$	$2\,1/2$	$1\,15/32$	$27/32$
$1\,5/8$	$2\,9/16$	$2\,61/64$	$1\,19/32$	$29/32$
$1\,3/4$	$2\,3/4$	$3\,11/64$	$1\,23/32$	$31/32$
$1\,7/8$	$2\,15/16$	$3\,25/64$	$1\,27/32$	$1\,1/32$
2	$3\,1/8$	$3\,39/64$	$1\,31/32$	$1\,3/32$

Nominal Dimensions of Square Nuts and Heavy Square Nuts

SQUARE NUTS   HEAVY SQUARE NUTS

| | SQUARE NUTS | | HEAVY SQUARE NUTS | | | |
| | Width Across Flats | | Width Across Corners | | Thickness | |
Nominal Size	Regular F	Heavy F	Regular G	Heavy G	Regular H	Heavy H
$1/4$	$7/16$	$1/2$	$5/8$	$45/64$	$7/32$	$1/4$
$5/16$	$9/16$	$9/16$	$51/64$	$51/64$	$17/64$	$5/16$
$3/8$	$5/8$	$11/16$	$57/64$	$31/32$	$21/64$	$3/8$
$7/16$	$3/4$	$3/4$	$1\,1/16$	$1\,1/16$	$3/8$	$7/16$
$1/2$	$13/16$	$7/8$	$1\,5/32$	$1\,15/64$	$7/16$	$1/2$
$5/8$	1	$1\,1/16$	$1\,27/64$	$1\,1/2$	$35/64$	$5/8$
$3/4$	$1\,1/8$	$1\,1/4$	$1\,19/32$	$1\,49/64$	$21/32$	$3/4$
$7/8$	$1\,9/16$	$1\,7/16$	$1\,55/64$	$2\,1/32$	$49/64$	$7/8$
1	$1\,1/2$	$1\,5/8$	$2\,1/8$	$2\,19/64$	$7/8$	1
$1\,1/8$	$1\,11/16$	$1\,13/16$	$2\,25/64$	$2\,9/16$	1	$1\,1/8$
$1\,1/4$	$1\,7/8$	2	$2\,21/32$	$2\,53/64$	$1\,3/32$	$1\,1/4$
$1\,3/8$	$2\,1/16$	$2\,3/16$	$2\,59/64$	$3\,3/32$	$1\,13/64$	$1\,3/8$
$1\,1/2$	$2\,1/4$	$2\,3/8$	$3\,3/16$	$3\,23/64$	$1\,5/16$	$1\,1/2$

Nominal Dimensions of Lag Screws

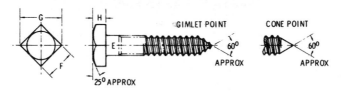

Nominal Size E	Width Across Flats F	Width Across Corners G	Head Height H
	$^9/_{32}$	$^{19}/_{64}$	$^1/_8$
$^1/_4$	$^3/_8$	$^{17}/_{32}$	$^{11}/_{64}$
$^5/_{16}$	$^1/_2$	$^{45}/_{64}$	$^{13}/_{64}$
$^3/_8$	$^9/_{16}$	$^{11}/_{16}$	$^1/_4$
$^7/_{16}$	$^5/_8$	$^{57}/_{64}$	$^{19}/_{64}$
$^1/_2$	$^3/_4$	$1\,^1/_{16}$	$^{21}/_{64}$
$^5/_8$	$^{15}/_{16}$	$1\,^{21}/_{64}$	$^{27}/_{64}$
$^3/_4$	$1\,^1/_8$	$1\,^{19}/_{32}$	$^1/_2$
$^7/_8$	$1\,^5/_{16}$	$1\,^{55}/_{64}$	$^{19}/_{32}$
1	$1\,^1/_2$	$2\,^1/_8$	$^{21}/_{32}$
$1\,^1/_8$	$1\,^{11}/_{16}$	$2\,^{25}/_{64}$	$^3/_4$
$1\,^1/_4$	$1\,^7/_8$	$2\,^{21}/_{32}$	$^{27}/_{32}$

American Standard Machine Screws*

Heads may be slotted or recessed

ROUND HEAD   FLAT HEAD   FILLISTER HEAD   OVAL HEAD   THRUSS HEAD

Nominal diam.	Round head		Flat head	Fillister head			Oval head		Truss head	
	A	H	A	A	H	O	A	C	A	H
0	0.113	0.053	0.119	0.096	0.045	0.059	0.119	0.021		
1	0.138	0.061	0.146	0.118	0.053	0.071	0.146	0.025	0.194	0.053
2	0.162	0.069	0.172	0.140	0.062	0.083	0.172	0.029	0.226	0.061
3	0.187	0.078	0.199	0.161	0.070	0.095	0.199	0.033	0.257	0.069
4	0.211	0.086	0.225	0.183	0.079	0.107	0.225	0.037	0.289	0.078
5	0.236	0.095	0.252	0.205	0.088	0.120	0.252	0.041	0.321	0.086
6	0.260	0.103	0.279	0.226	0.096	0.132	0.279	0.045	0.352	0.094
8	0.309	0.120	0.332	0.270	0.113	0.156	0.332	0.052	0.384	0.102
10	0.359	0.137	0.385	0.313	0.130	0.180	0.385	0.060	0.448	0.118
12	0.408	0.153	0.438	0.357	0.148	0.205	0.438	0.068	0.511	0.134
$\frac{1}{4}$	0.472	0.175	0.507	0.414	0.170	0.237	0.507	0.079	0.573	0.150
$\frac{5}{16}$	0.590	0.216	0.635	0.518	0.211	0.295	0.635	0.099	0.698	0.183
$\frac{3}{8}$	0.708	0.256	0.762	0.622	0.253	0.355	0.762	0.117	0.823	0.215
$\frac{7}{16}$	0.750	0.328	0.812	0.625	0.265	0.368	0.812	0.122	0.948	0.248
$\frac{1}{2}$	0.813	0.355	0.875	0.750	0.297	0.412	0.875	0.131	1.073	0.280
$\frac{9}{16}$	0.938	0.410	1.000	0.812	0.336	0.466	1.000	0.150	1.198	0.312
$\frac{5}{8}$	1.000	0.438	1.125	0.875	0.375	0.521	1.125	0.169	1.323	0.345
$\frac{3}{4}$	1.250	0.547	1.375	1.000	0.441	0.612	1.375	0.206	1.573	0.410

American Standard Machine Screws* (Cont'd)

Heads may be slotted or recessed

BINDING HEAD   PAN HEAD   PAN HEAD (RECESSED)   HEXAGON HEAD   100° FLAT HEAD

Nominal diam.	Binding head				Pan head			Hexagon head		100° flat head
	A	O	F	U	A	H	O	A	H	A
2	0.181	0.046	0.018	0.141	0.167	0.053	0.062	0.125	0.050	
3	0.208	0.054	0.022	0.162	0.193	0.060	0.071	0.187	0.055	
4	0.235	0.063	0.025	0.184	0.219	0.068	0.080	0.187	0.060	0.225
5	0.263	0.071	0.029	0.205	0.245	0.075	0.089	0.187	0.070	
6	0.290	0.080	0.032	0.226	0.270	0.082	0.097	0.250	0.080	0.279
8	0.344	0.097	0.039	0.269	0.322	0.096	0.115	0.250	0.110	0.332
10	0.399	0.114	0.045	0.312	0.373	0.110	0.133	0.312	0.120	0.385
12	0.454	0.130	0.052	0.354	0.425	0.125	0.151	0.312	0.155	
¼	0.513	0.153	0.061	0.410	0.492	0.144	0.175	0.375	0.190	0.507
⁵⁄₁₆	0.641	0.193	0.077	0.513	0.615	0.178	0.218	0.500	0.230	0.635
³⁄₈	0.769	0.234	0.094	0.615	0.740	0.212	0.261	0.562	0.295	0.762

*ANSI B18.6—1972. Dimensions given are maximum values, all in inches. Thread length: screws 2 in. long or less, thread entire length; screws over 2 in. long, thread length $1 = 1\frac{3}{4}$ in. Threads are coarse or fine series, class 2. Heads may be slotted or recessed as specified, excepting hexagon form, which is plain or may be slotted if so specified. Slot and recess proportions vary with size of fastener; draw to look well.

American Standard Hexagon Socket,* Slotted Headless,† and Square-head‡ Setscrews

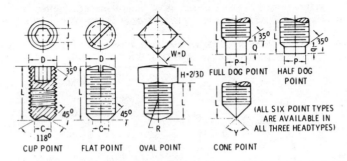

CUP POINT   FLAT POINT   OVAL POINT   CONE POINT

FULL DOG POINT   HALF DOG POINT

(ALL SIX POINT TYPES ARE AVAILABLE IN ALL THREE HEADTYPES)

Diam. D	Cup and flat-point diam. C	Oval-point radius R	Cone-point angle Y		Full and half dog points			Socket width J
			118° for these lengths and shorter	90° for these lengths and longer	Diam. P	Length Full Q	Length Half q	
5	$^1/_{16}$	$^3/_{32}$	$^1/_8$	$^3/_{16}$	0.083	0.06	0.03	$^1/_{16}$
6	0.069	$^7/_{64}$	$^1/_8$	$^3/_{16}$	0.092	0.07	0.03	$^1/_{16}$
8	$^5/_{64}$	$^1/_8$	$^3/_{16}$	$^1/_4$	0.109	0.08	0.04	$^5/_{64}$
10	$^3/_{32}$	$^9/_{64}$	$^3/_{16}$	$^1/_4$	0.127	0.09	0.04	$^3/_{32}$
12	$^7/_{64}$	$^5/_{32}$	$^3/_{16}$	$^1/_4$	0.144	0.11	0.06	$^3/_{32}$
$^1/_4$	$^1/_8$	$^3/_{16}$	$^1/_4$	$^5/_{16}$	$^5/_{32}$	$^1/_8$	$^1/_{16}$	$^1/_8$
$^5/_{16}$	$^{11}/_{64}$	$^{15}/_{64}$	$^5/_{16}$	$^3/_8$	$^{13}/_{64}$	$^5/_{32}$	$^5/_{64}$	$^5/_{32}$
$^3/_8$	$^{13}/_{64}$	$^9/_{32}$	$^3/_8$	$^7/_{16}$	$^1/_4$	$^3/_{16}$	$^3/_{32}$	$^3/_{16}$
$^7/_{16}$	$^{15}/_{64}$	$^{21}/_{64}$	$^7/_{16}$	$^1/_2$	$^{19}/_{64}$	$^7/_{32}$	$^7/_{64}$	$^7/_{32}$
$^1/_2$	$^9/_{32}$	$^3/_8$	$^1/_2$	$^9/_{16}$	$^{11}/_{32}$	$^1/_4$	$^1/_8$	$^1/_4$
$^9/_{16}$	$^5/_{16}$	$^{27}/_{64}$	$^9/_{16}$	$^5/_8$	$^{25}/_{64}$	$^9/_{32}$	$^9/_{64}$	$^1/_4$
$^5/_8$	$^{23}/_{64}$	$^{15}/_{32}$	$^5/_8$	$^3/_4$	$^{15}/_{32}$	$^5/_{16}$	$^5/_{32}$	$^5/_{16}$
$^3/_4$	$^7/_{16}$	$^9/_{16}$	$^3/_4$	$^7/_8$	$^9/_{16}$	$^3/_8$	$^3/_{16}$	$^3/_8$
$^7/_8$	$^{33}/_{64}$	$^{21}/_{32}$	$^7/_8$	1	$^{21}/_{32}$	$^7/_{16}$	$^7/_{32}$	$^1/_2$
1	$^{19}/_{32}$	$^3/_4$	1	$1^1/_8$	$^3/_4$	$^1/_2$	$^1/_4$	$^9/_{16}$
$1^1/_8$	$^{43}/_{64}$	$^{27}/_{32}$	$1^1/_8$	$1^1/_4$	$^{27}/_{32}$	$^9/_{16}$	$^9/_{32}$	$^9/_{16}$
$1^1/_4$	$^3/_4$	$^{15}/_{16}$	$1^1/_4$	$1^1/_2$	$^{15}/_{16}$	$^5/_8$	$^5/_{16}$	$^5/_8$
$1^3/_8$	$^{53}/_{64}$	$1^1/_{32}$	$1^3/_8$	$1^5/_8$	$1^1/_{32}$	$^{11}/_{16}$	$^{11}/_{32}$	$^5/_8$
$1^1/_2$	$^{29}/_{32}$	$1^1/_8$	$1^1/_2$	$1^3/_4$	$1^1/_8$	$^3/_4$	$^3/_8$	$^3/_4$
$1^3/_4$	$1^1/_{16}$	$1^5/_{16}$	$1^3/_4$	2	$1^5/_{16}$	$^7/_8$	$^7/_{16}$	1
2	$1^7/_{32}$	$1^1/_2$	2	$2^1/_4$	$1^1/_2$	1	$^1/_2$	1

*ANSI B18.3—1976. Dimensions are in inches. Threads coarse or fine, class 3A. Length increments; $^1/_4$ in. to $^5/_8$ in. by ($^1/_{16}$ in.); $^5/_8$ in. to 1 in. by ($^1/_8$ in.); 1 in. to

4 in. by ($\frac{1}{4}$ in.); 4 in. to 6 in. by ($\frac{1}{2}$ in.). Fractions in parentheses show length increments; for example, $\frac{5}{8}$ in. to 1 in. by ($\frac{1}{8}$ in.) includes the lengths $\frac{5}{8}$ in., $\frac{3}{4}$ in., $\frac{7}{8}$ in., and 1 in.

†ANSI B18.6.2—1972. Threads coarse or fine, class 2A. Slotted headless screws standardized in sizes No. 5 to $\frac{3}{4}$ in. only. Slot proportions vary with diameter. Draw to look well.

‡ANSI B18.6.2—1972. Threads coarse, fine, or 8-pitch, class 2A. Square-head setscrews standardized in sizes No. 10 to 1$\frac{1}{2}$ in. only.

American Standard Cap Screws[a]—Socket[b] and Slotted Heads[c]

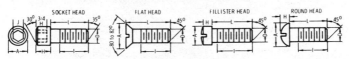

Nominal diam.	Socket head[d]			Flat head[e]	Filister head[e]		Round head[e]	
	A	H	J	A	A	H	A	H
0	0.096	0.060	0.050					
1	0.118	0.073	0.050					
2	0.140	0.086	1/16					
3	0.161	0.099	5/64					
4	0.183	0.112	5/64					
5	0.205	0.125	3/32					
6	0.226	0.138	3/32					
8	0.270	0.164	1/8					
10	5/16	0.190	5/32					
12	11/32	0.216	5/32					
1/4	3/8	1/4	3/16	1/2	3/8	11/64	7/16	3/16
5/16	7/16	5/16	7/32	5/8	7/16	13/64	9/16	15/64
3/8	9/16	3/8	5/16	3/4	9/16	1/4	5/8	17/64
7/16	5/8	7/16	5/16	13/16	5/8	19/64	3/4	5/16
1/2	3/4	1/2	3/8	7/8	3/4	21/64	13/16	11/32
9/16	13/16	9/16	3/8	1	17/16	7/8	15/16	13/32
5/8	7/8	5/8	1/2	1 1/8	7/8	27/64	1	7/16
3/4	1	3/4	9/16	1 3/8	1	1/2	1 1/4	17/32
7/8	1 1/8	7/8	9/16	1 5/8	1 1/8	19/32		
1	1 5/16	1	5/8	1 7/8	1 5/16	21/32		
1 1/8	1 1/2	1 1/8	3/4					
1 1/4	1 3/4	1 1/4	3/4					
1 3/8	1 7/8	1 3/8	3/4					
1 1/2	2	1 1/2	1					

[a]Dimensions in inches.
[b]ANSI B18.3—1976.
[c]ANSI B18.6.2—1972.
[d]Thread coarse or fine, class 3A. Thread length l: coarse thread, $2D + 1/2$ in.; fine thread, $1\frac{1}{2}D + 1/2$ in.
[e]Thread coarse, fine, or 8-pitch, class 2A. Thread length l: $2D + 1/4$ in.
Slot proportions vary with size of screw; draw to look well. All body-length increments for screw lengths 1/4 in. to 1 in. = 1/8 in., for screw lengths 1 in. to 4 in. = 1/4 in., for screw lengths 4 in. to 6 in. = 1/2 in.

English to Metric Conversion Table

Decimals to Millimeters				Fractions to Decimals to Millimeters					
Decimal	mm	Decimal	mm	Fraction	Decimal	mm	Fraction	Decimal	mm
0.001	0.0254	0.200	5.0800	1/64	0.0156	0.3969	17/64	0.2656	6.7469
0.002	0.0508	0.210	5.3340	1/32	0.0312	0.7938	9/32	0.2812	7.1438
0.003	0.0762	0.220	5.5880	3/64	0.0469	1.1906	19/64	0.2969	7.5406
0.004	0.1016	0.230	5.8420						
0.005	0.1270	0.240	6.0960						
0.006	0.1524	0.250	6.3500	1/16	0.0625	1.5875	5/16	0.3125	7.9375
0.007	0.1778	0.260	6.6040						
0.008	0.2032	0.270	6.8580						
0.009	0.2286	0.280	7.1120	5/64	0.0781	1.9844	21/64	0.3281	8.3344
		0.290	7.3660	3/32	0.0938	2.3812	11/32	0.3438	8.7312
				7/64	0.1094	2.7781	23/64	0.3594	9.1281
0.010	0.2540	0.300	7.6200						
0.020	0.5080	0.310	7.8740						
0.030	0.7620	0.320	8.1280	1/8	0.1250	3.1750	3/8	0.3750	9.5250
0.040	1.0160	0.330	8.3820						
0.050	1.2700	0.340	8.6360						
0.060	1.5240	0.350	8.8900	9/64	0.1406	3.5719	25/64	0.3906	9.9219
0.070	1.7780	0.360	9.1440	5/32	0.1562	3.9688	13/32	0.4062	10.3188
0.080	2.0320	0.370	9.3980	11/64	0.1719	4.3656	27/64	0.4219	10.7156
0.090	2.2860	0.380	9.6520						
		0.390	9.9060	3/16	0.1875	4.7625	7/16	0.4375	11.1125
0.100	2.5400	0.400	10.1600						
0.110	2.7940	0.410	10.4140						
0.120	3.0480	0.420	10.6680	13/64	0.2031	5.1594	29/64	0.4531	11.5094
0.130	3.3020	0.430	10.9220	7/32	0.2188	5.5562	15/32	0.4688	11.9062
0.140	3.5560	0.440	11.1760	15/64	0.2344	5.9531	31/64	0.4844	12.3031
0.150	3.8100	0.450	11.4300						
0.160	4.0640	0.460	11.6840						
0.170	4.3180	0.470	11.9380	1/4	0.2500	6.3500	1/2	0.5000	12.7000
0.180	4.5720	0.480	12.1920						
0.190	4.8260	0.490	12.4460						

English to Metric Conversion Table (Cont'd)

Decimals to Millimeters

Decimal	mm	Decimal	mm
0.500	12.7000	0.750	19.0500
0.510	12.9540	0.760	19.3040
0.520	13.2080	0.770	19.5580
0.530	13.4620	0.780	19.8120
0.540	13.7160	0.790	20.0660
0.550	13.9700		
0.560	14.2240		
0.570	14.4780	0.800	20.3200
0.580	14.7320	0.810	20.5740
0.590	14.9860	0.820	20.8280
		0.830	21.0820
		0.840	21.3360
		0.850	21.5900
0.600	15.2400	0.860	21.8440
0.610	15.4940	0.870	22.0980
0.620	15.7480	0.880	22.3520
0.630	16.0020	0.890	22.6060
0.640	16.2560		
0.650	16.5100		
0.660	16.7640		
0.670	17.0180	0.900	22.8600
0.680	17.2720	0.910	23.1140
0.690	17.5260	0.920	23.3680
		0.930	23.6220
		0.940	23.8760
		0.950	24.1300
0.700	17.7800	0.960	24.3840
0.710	18.0340	0.970	24.6380
0.720	18.2880	0.980	24.8920
0.730	18.5420	0.990	25.1460
0.740	18.7960	1.000	25.4000

Fractions to Decimals to Millimeters

Fraction	Decimal	mm	Fraction	Decimal	mm
33/64	0.5156	13.0969	49/64	0.7656	19.4469
17/32	0.5312	13.4938	25/32	0.7812	19.8438
35/64	0.5469	13.8906	51/64	0.7969	20.2406
9/16	0.5625	14.2875	13/16	0.8125	20.6375
37/64	0.5781	14.6844	53/64	0.8281	21.0344
19/32	0.5938	15.0812	27/32	0.8438	21.4312
39/64	0.6094	15.4781	55/64	0.8594	21.8281
5/8	0.6250	15.8750	7/8	0.8750	22.2250
41/64	0.6406	16.2719	57/64	0.8906	22.6219
21/32	0.6562	16.6688	29/32	0.9062	23.0188
43/64	0.6719	17.0656	59/64	0.9219	23.4156
11/16	0.6875	17.4625	15/16	0.9375	23.8125
45/64	0.7031	17.8594	61/64	0.9531	24.2094
23/32	0.7188	18.2562	31/32	0.9688	24.6062
47/64	0.7344	18.6531	63/64	0.9844	25.0031
3/4	0.7500	19.0500	1	1.0000	25.4000

Decimal Equivalents, Squares, Cubes, Square and Cube Roots, Circumferences and Areas of Circles, from $\frac{1}{64}$ to $\frac{5}{8}$ Inch

Fraction	Dec. Equiv.	Square	Square Root	Cube	Cube Root	Circle* Circum.	Circle* Area
$\frac{1}{64}$	.015625	.0002441	.125	.000003815	.25	.04909	.000192
$\frac{1}{32}$	.03125	.0009766	.176777	.000030518	.31498	.09817	.000767
$\frac{3}{64}$	.046875	.0021973	.216506	.000102997	.36056	.14726	.001726
$\frac{1}{16}$	.0625	.0039063	.25	.00024414	.39685	.19635	.003068
$\frac{5}{64}$	.078125	.0061035	.279508	.00047684	.42749	.24544	.004794
$\frac{3}{32}$	.09375	.0087891	.306186	.00082397	.45428	.29452	.006903
$\frac{7}{64}$	.109375	.0119629	.330719	.0013084	.47823	.34361	.009396
$\frac{1}{8}$	.125	.015625	.353553	.0019531	.5	.39270	.012272
$\frac{9}{64}$	.140625	.0197754	.375	.0027809	.52002	.44179	.015532
$\frac{5}{32}$	.15625	.0244141	.395285	.0038147	.53861	.49087	.019175
$\frac{11}{64}$	.171875	.0295410	.414578	.0050774	.55600	.53996	.023201
$\frac{3}{16}$	.1875	.0351563	.433013	.0065918	.57236	.58905	.027611
$\frac{13}{64}$	.203125	.0412598	.450694	.0083809	.58783	.63814	.032405
$\frac{7}{32}$	.21875	.0478516	.467707	.010468	.60254	.68722	.037583
$\frac{15}{64}$	.234375	.0549316	.484123	.012875	.61655	.73631	.043143
$\frac{1}{4}$	.25	.0625	.5	.015625	.62996	.78540	.049087
$\frac{17}{64}$	.265625	.0705566	.515388	.018742	.64282	.83449	.055415
$\frac{9}{32}$	.28125	.0791016	.530330	.022247	.65519	.88357	.062126
$\frac{19}{64}$	.296875	.0881348	.544862	.026165	.66710	.93266	.069221
$\frac{5}{16}$	.3125	.0976562	.559017	.030518	.67860	.98175	.076699
$\frac{21}{64}$	.328125	.107666	.572822	.035328	.68973	1.03084	.084561
$\frac{11}{32}$	.34375	.118164	.586302	.040619	.70051	1.07992	.092806
$\frac{23}{64}$	.359375	.129150	.599479	.046413	.71097	1.12901	.101434
$\frac{3}{8}$	.375	.140625	.612372	.052734	.72112	1.17810	.110445
$\frac{25}{64}$	.390625	.1525879	.625	.059605	.73100	1.22718	.119842
$\frac{13}{32}$	.40625	.1650391	.637377	.067047	.74062	1.27627	.129621
$\frac{27}{64}$	.421875	.1779785	.649519	.075085	.75	1.32536	.139784
$\frac{7}{16}$	.4375	.1914063	.661438	.083740	.75915	1.37445	.150330
$\frac{29}{64}$	.453125	.2053223	.673146	.093037	.76808	1.42353	.161260
$\frac{15}{32}$	.46875	.2197266	.684653	.102997	.77681	1.47262	.172573
$\frac{31}{64}$	.484375	.2346191	.695971	.113644	.78535	1.52171	.184269
$\frac{1}{2}$	.5	.25	.707107	.125	.79370	1.57080	.196350
$\frac{33}{64}$	.515625	.265869	.718070	.137089	.80188	1.61988	.208813
$\frac{17}{32}$	.53125	.282227	.728869	.149933	.80990	1.66897	.221660
$\frac{35}{64}$	.546875	.299072	.739510	.163555	.81777	1.71806	.234891
$\frac{9}{16}$	.5625	.316406	.75	.177979	.82548	1.76715	.248505
$\frac{37}{64}$	.578125	.334229	.760345	.193226	.83306	1.81623	.262502
$\frac{19}{32}$	.59375	.352539	.770552	.209320	.84049	1.86532	.276884
$\frac{39}{64}$	.609375	.371338	.780625	.226284	.84780	1.91441	.291648
$\frac{5}{8}$	.625	.390625	.790569	.244141	.85499	1.96350	.306796

*Fraction represents diameter.

Decimal Equivalents, Squares, Cubes, Square and Cube Roots, Circumferences and Areas of Circles, from $\frac{41}{64}$ to 1 Inch

Fraction	Dec. Equiv.	Square	Square Root	Cube	Cube Root	Circle*	
						Circum.	Area
$\frac{41}{64}$	.640625	.410400	.800391	.262913	.86205	2.01258	.322328
$\frac{21}{32}$	.65625	.430664	.810093	.282623	.86901	2.06167	.338243
$\frac{43}{64}$	.671875	.451416	.819680	.303295	.87585	2.11076	.354541
$\frac{11}{16}$	.6875	.472656	.829156	.324951	.88259	2.15984	.371223
$\frac{45}{64}$	.703125	.494385	.838525	.347614	.88922	2.20893	.388289
$\frac{23}{32}$	.71875	.516602	.847791	.371307	.89576	2.25802	.405737
$\frac{47}{64}$	.734375	.539307	.856957	.396053	.90221	2.30711	.423570
$\frac{3}{4}$	.75	.5625	.866025	.421875	.90856	2.35619	.441786
$\frac{49}{64}$	.765625	.586182	.875	.448795	.91483	2.40528	.460386
$\frac{25}{32}$	.78125	.610352	.883883	.476837	.92101	2.45437	.479369
$\frac{51}{64}$	.796875	.635010	.892679	.506023	.92711	2.50346	.498736
$\frac{13}{16}$	.8125	.660156	.901388	.536377	.93313	2.55254	.518486
$\frac{53}{64}$	.828125	.685791	.910014	.567921	.93907	2.60163	.538619
$\frac{27}{32}$	.84375	.711914	.918559	.600677	.94494	2.65072	.559136
$\frac{55}{64}$	.859375	.738525	.927024	.634670	.95074	2.69981	.580036
$\frac{7}{8}$	.875	.765625	.935414	.669922	.95647	2.74889	.601320
$\frac{57}{64}$	.890625	.793213	.943729	.706455	.96213	2.79798	.622988
$\frac{29}{32}$	.90625	.821289	.951972	.744293	.96772	2.84707	.645039
$\frac{59}{64}$	.921875	.849854	.960143	.783459	.97325	2.89616	.667473
$\frac{15}{16}$	.9375	.878906	.968246	.823975	.97872	2.94524	.690291
$\frac{61}{64}$	.953125	.908447	.976281	.865864	.98412	2.99433	.713493
$\frac{31}{32}$	.96875	.938477	.984251	.909149	.98947	3.04342	.737078
$\frac{63}{64}$	.984375	.968994	.992157	.953854	.99476	3.09251	.761046
1	1	1	1	1	1	3.14159	.785398

*Fraction represents diameter.

Screw Threads and Tap Drill Sizes

Size of Tap	NC or ASME SPECIAL MACHINE SCREWS Threads per Inch	Tap Drill	Body Drill	Size of Tap	NF or ASME SPECIAL MACHINE SCREWS Threads per Inch	Tap Drill	Body Drill
1	64	53	48	2	64	50	44
2	56	50	44	3	56	45	39
3	48	47	39	4	48	42	33
4	40	43	33	5	44	37	$1/8$
5	40	38	$1/8$	6	40	33	28
6	32	36	28	8	36	29	19
8	32	29	19	10	32	21	11
10	24	25	11	*10	30	22	11
12	24	16	$7/32$	12	28	14	$7/32$

*A S M E Only

Size of Tap	AMERICAN STANDARD TAPER PIPE THREADS Threads per Inch	Tap Drill	Size of Tap	NF or SAE STANDARD SCREWS Threads per Inch	Tap Drill
$1/8$	27	$11/32$	$1/4$	28	3
$1/4$	18	$7/16$	$5/16$	24	1
$3/8$	18	$19/32$	$3/8$	24	0
$1/2$	14	$23/32$	$7/16$	20	$25/64$
$3/4$	14	$15/16$	$1/2$	20	$25/64$
1	$11 1/2$	$1 5/32$	$9/16$	18	$33/64$
$1 1/4$	$11 1/2$	$1 1/2$	$5/8$	18	$37/64$
$1 1/2$	$11 1/2$	$1 23/32$	$11/16$	16	$5/8$
2	$11 1/2$	$2 3/16$	$3/4$	16	$11/16$
$2 1/2$	8	$2 5/8$	$7/8$	14	$13/16$
3	8	$3 1/4$	1	14	$15/16$
			$1 1/8$	12	$1 3/64$

*Tap Drills allow approx. 75% full thread.

Number and Letter Sizes of Drills with Decimal Equivalents

Sizes starting with No. 80 and going up to 1 inch. This table is useful for quickly determining the nearest drill size for any decimal, for root diameters, body drills, etc.

Drill No.	Frac.	Deci.	Drill No.	Frac.	Deci.	Drill No.	Frac.	Deci.	Drill No.	Frac.	Deci.
80	—	.0135	42	—	.0935	7	—	.201	X	—	.397
79	—	.0145	—	3/32	.0938	—	13/64	.203	Y	—	.404
—	1/64	.0156				6	—	.204		13/32	.406
78	—	.0160	41	—	.0960	5	—	.206	Z	—	.413
77	—	.0180	40	—	.0980	4	—	.209	—	27/64	.422
			39	—	.0995				—	7/16	.438
76	—	.0200	38	—	.1015	3	—	.213	—	29/64	.453
75	—	.0210	37	—	.1040	—	7/32	.219			
74	—	.0225				2	—	.221			
73	—	.0240	36	—	.1065	1	—	.228	—	15/32	.469
72	—	.0250	—	7/64	.1094	A	—	.234	—	31/64	.484
			35	—	.1100				—	1/2	.500
71	—	.0260	34	—	.1110	—	15/64	.234	—	33/64	.516
70	—	.0280	33	—	.1130	B	—	.238	—	17/32	.531
69	—	.0292				C	—	.242			
68	—	.0310	32	—	.116	D	—	.246	—	35/64	.547
—	1/32	.0313	31	—	.120	—	1/4	.250	—	9/16	.562
			—	1/8	.125				—	37/64	.578
67	—	.0320	30	—	.129	E	—	.250	—	19/32	.594
66	—	.0330	29	—	.136	F	—	.257	—	39/64	.609
65	—	.0350				G	—	.261			
64	—	.0360	—	9/64	.140	—	17/64	.266	—	5/8	.625
63	—	.0370	28	—	.141	H	—	.266	—	41/64	.641
			27	—	.144				—	21/32	.656
62	—	.0380	26	—	.147	I	—	.272	—	43/64	.672
61	—	.0390	25	—	.150	J	—	.277	—	11/16	.688
60	—	.0400				—	9/32	.281			
59	—	.0410	24	—	.152	K	—	.281	—	45/64	.703
58	—	.0420	23	—	.154	L	—	.290	—	23/32	.719
			—	5/32	.156				—	47/64	.734
57	—	.0430	22	—	.157	M	—	.295	—	3/4	.750
56	—	.0465	21	—	.159	—	19/64	.297	—	49/64	.766
—	3/64	.0469				N	—	.302			
55	—	.0520				—	5/16	.313	—	25/32	.781
54	—	.0550	20	—	.161	O	—	.316	—	51/64	.797
			19	—	.166				—	13/16	.813
53	—	.0595	18	—	.170	P	—	.323	—	53/64	.828
—	1/16	.0625	—	11/64	.172	—	21/64	.328	—	27/32	.844
52	—	.0635	17	—	.173	Q	—	.332			
51	—	.0670				R	—	.339			
50	—	.0700	16	—	.177	—	11/32	.344	—	55/64	.859
			15	—	.180				—	7/8	.875
49	—	.0730	14	—	.182	S	—	.348	—	57/64	.891
48	—	.0760	13	—	.185	T	—	.358	—	29/32	.906
—	5/64	.0781	—	3/16	.188	—	23/64	.359	—	59/64	.922
47	—	.0785				U	—	.368			
46	—	.0810	12	—	.189	—	3/8	.375	—	15/16	.938
			11	—	.191				—	61/64	.953
45	—	.0820	10	—	.194	V	—	.377	—	31/32	.969
44	—	.0860	9	—	.196	W	—	.386	—	63/64	.984
43	—	.0890	8	—	.199	—	25/64	.391	—	1	1.000

Index

The Audel® Mail Order Bookstore

Here's an opportunity to order the valuable books you may have missed before and to build your own personal, comprehensive library of Audel books. You can choose from an extensive selection of technical guides and reference books. They will provide access to the same sources the experts use, put all the answers at your fingertips, and give you the know-how to complete even the most complicated building or repairing job, in the same professional way.

Each volume:
- **Fully illustrated**
- **Packed with up-to-date facts and figures**
- **Completely indexed for easy reference**

APPLIANCES

HOME APPLIANCE SERVICING, 3rd Edition
A practical book for electric & gas servicemen, mechanics & dealers. Covers the principles, servicing, and repairing of home appliances. 592 pages; $5\frac{1}{2}\times8\frac{1}{4}$; hardbound. **Price: $12.95**

REFRIGERATION AND AIR CONDITIONING LIBRARY—2 Vols. Price: $21.95

REFRIGERATION: HOME AND COMMERCIAL
Covers the whole realm of refrigeration equipment from fractional-horsepower water coolers, through domestic refrigerators to multi-ton commercial installations. 656 pages; $5\frac{1}{2}\times8\frac{1}{4}$; hardbound. **Price: $12.95**

AIR CONDITIONING: HOME AND COMMERCIAL
A concise collection of basic information, tables, and charts for those interested in understanding troubleshooting, and repairing home air conditioners and commercial installations. 464 pages; $5\frac{1}{2}\times8\frac{1}{4}$; hardbound. **Price: $10.95**

OIL BURNERS, 3rd Edition
Provides complete information on all types of oil burners and associated equipment. Discusses burners—blowers—ignition transformers—electrodes—nozzles—fuel pumps—filters—controls. Installation and maintenance are stressed. 320 pages; $5\frac{1}{2}\times8\frac{1}{4}$; hardbound. **Price: $9.95**

Use the order coupon on the back of this book.
All prices are subject to change without notice.

AUTOMOTIVE

AUTOMOBILE REPAIR GUIDE, 4th Edition
A practical reference for auto mechanics, servicemen, trainees, and owners. Explains theory, construction, and servicing of modern domestic motorcars. 800 pages; 5½×8¼; hardbound. **Price: $14.95**

AUTOMOTIVE AIR CONDITIONING
You can easily perform most all service procedures you've been paying for in the past. This book covers the systems built by the major manufacturers, even after-market installations. Contents: introduction—refrigerant—tools—air conditioning circuit—general service procedures—electrical systems—the cooling system—system diagnosis—electrical diagnosis—troubleshooting. 232 pages; 5½×8¼; softcover. **Price: $7.95**

DIESEL ENGINE MANUAL
A practical guide covering the theory, operation and maintenance of modern diesel engines. Explains diesel principles—valves—timing—fuel pumps—pistons and rings—cylinders—lubrication—cooling system—fuel oil and more. 480 pages; 5½×8¼; hardbound **Price: $12.95**

GAS ENGINE MANUAL, 2nd Edition
A completely practical book covering the construction, operation, and repair of all types of modern gas engines. 400 pages; 5½×8¼; hardbound. **Price: $9.95.**

BUILDING AND MAINTENANCE

ANSWERS ON BLUEPRINT READING, 3rd Edition
Covers all types of blueprint reading for mechanics and builders. This book reveals the secret language of blueprints, step by step in easy stages. 312 pages; 5½×8¼; hardbound. **Price: $9.95**

BUILDING MAINTENANCE, 2nd Edition
Covers all the practical aspects of building maintenace. Painting and decorating; plumbing and pipe fitting; carpentry; heating maintenace; custodial practices and more. (A book for building owners, managers, and maintenance personnel.) 384 pages; 5½×8¼; hardbound. **Price: $9.95**

COMPLETE BUILDING CONSTRUCTION
At last—a one volume instruction manual to show you how to construct a frame or brick building from the footings to the ridge. Build your own garage, tool shed, other outbuildings—even your own house or place of business. Building construction tells you how to lay out the building and excavation lines on the lot; how to make concrete forms and pour the footings and foundation; how to make concrete slabs, walks, and driveways; how to lay concrete block, brick and tile; how to build your own fireplace and chimney. It's one of the newest Audel books, clearly written by experts in each field and ready to help you every step of the way. 800 pages; 5½×8¼; hardbound. **Price: $19.95**

Use the order coupon on the back of this book.
All prices are subject to change without notice.

GARDENING & LANDSCAPING
A comprehensive guide for homeowners and for industrial, municipal, and estate grounds-keepers. Gives information on proper care of annual and perennial flowers; various house plants; greenhouse design and construction; insect and rodent controls; and more. 384 pages; 5½×8¼ hardbound. **Price: $9.95**

CARPENTERS & BUILDERS LIBRARY, 4th Edition (4 Vols.)
A practical, illustrated trade assistant on modern construction for carpenters, builders, and all woodworkers. Explains in practical, concise language and illustrations all the principles, advances, and shortcuts based on modern practice. How to calculate various jobs. **Price: $39.95**

Vol. 1—Tools, steel square, saw filing, joinery cabinets. 384 pages; 5½×8¼; hardbound. **Price: $10.95**
Vol. 2—Mathematics, plans, specifications, estimates. 304 pages; 5½×8¼; hardbound. **Price: $10.95**
Vol. 3—House and roof framing, layout foundations. 304 pages; 5½×8¼; hardbound. **Price: $10.95**
Vol. 4—Doors, windows, stairs, millwork, painting. 368 pages; 5½×8¼; hardbound. **Price: $10.95**

CARPENTRY AND BUILDING
Answers to the problems encountered in today's building trades. The actual questions asked of an architect by carpenters and builders are answered in this book. 448 pages; 5½×8¼; hardbound. **Price: $10.95**

WOOD STOVE HANDBOOK
The wood stove handbook shows how wood burned in a modern wood stove offers an immediate, practical, low-cost method of full-time or part-time home heating. The book points out that wood is plentiful, low in cost (sometimes free), and nonpolluting, especially when burned in one of the newer and more efficent stoves. In this book, you will learn about the nature of heat and its control, what happens inside and outside a stove, how to have a safe and efficient chimney, and how to install a modern wood burning stove. You will learn about the differnt types of firewood and how to get it, cut it, split it and store it. 128 pages; 8½×11; softcover. **Price: $7.95**

HEATING, VENTILATING, AND AIR CONDITIONING LIBRARY (3 Vols.)
This three-volume set covers all types of furnaces, ductwork, air conditioners, heat pumps, radiant heaters, and water heaters, including swimming-pool heating systems. **Price: $38.95**

Volume 1
Partial Contents: Heating Fundamentals . . . Insulation Principles . . . Heating Fuels . . . Electric Heating System . . . Furnace Fundamentals . . . Gas-Fired Furnaces . . . Oil-Fired Furnaces . . . Coal-Fired Furnaces . . . Electric Furnaces. 614 pages; 5½×8¼; hardbound. **Price: $13.95**

Volume 2
Partial Contents: Oil Burners . . . Gas Burners . . . Thermostats and Humidistats . . . Gas and Oil Controls . . . Pipes, Pipe Fitting, and Piping Details . . . Valves and Valve Installations. 560 pages; 5½×8¼; hardbound. **Price $13.95**

Volume 3
Partial Contents: Radiant Heating . . . Radiators, Convectors, and Unit Heaters . . . Stoves, Fireplaces, and Chimneys . . . Water Heaters and Other Appliances . . . Central Air Conditioning Systems . . . Humidifiers and Dehumidifiers. 544 pages; 5½×8¼; hardbound. **Price: $13.95**

HOME MAINTENANCE AND REPAIR: Walls, Ceilings, and Floors
Easy-to-follow instructions for sprucing up and repairing the walls, ceiling, and floors of your home. Covers nail pops, plaster repair, painting, paneling, ceiling and bathroom tile, and sound control. 80 pages; 8½×11; softcover. **Price: $6.95**

HOME PLUMBING HANDBOOK, 2nd Edition
A complete guide to home plumbing repair and installation. 200 pages; 8½×11; softcover. **Price: $7.95**

Use the order coupon on the back of this book.
All prices are subject to change without notice.

MASONS AND BUILDERS LIBRARY—2 Vols.

A practical, illustrated trade assistant on modern construction for bricklayers, stonemasons, cement workers, plasters, and tile setters. Explains all the principles, advances, and shortcuts based on modern practice—including how to figure and calculate various jobs. **Price $17.95**

> Vol. 1— Concrete Block, Tile, Terrazzo. 368 pages; 5½×8¼; hardbound. **Price: $9.95**
> Vol. 2—Bricklaying, Plastering, Rock Masonry, Clay Tile. 384 pages; 5½×8¼; hardbound. **Price: $9.95**

PLUMBERS AND PIPE FITTERS LIBRARY—3 Vols.

A practical, illustrated trade assistant and reference for master plumbers, journeymen and apprentice pipe fitters, gas fitters and helpers, builders, contractors, and engineers. Explains in simple language, illustrations, diagrams, charts, graphs, and pictures the principles of modern plumbing and pipe-fitting practices. **Price $29.95**

> Vol. 1—Materials, tools, roughing-in. 320 pages; 5½×8¼; hardbound. **Price: $10.95**
> Vol. 2—Welding, heating air-conditioning. 384 pages; 5½×8¼; hardbound. **Price: $10.95**
> Vol. 3—Water supply, drainage, calculations. 272 pages; 5½×8¼; hardbound. **Price: $10.95**

PLUMBERS HANDBOOK

A pocket manual providing reference material for plumbers and/or pipe fitters. General information sections contain data on cast-iron fittings, copper drainage fittings, plastic pipe, and repair of fixtures. 288 pages; 4×6 softcover. **Price: $9.95**

QUESTIONS AND ANSWERS FOR PLUMBERS EXAMINATIONS, 2nd Edition

Answers plumbers' questions about types of fixtures to use, size of pipe to install, design of systems, size and location of septic tank systems, and procedures used in installing material. 256 pages; 5½×8¼; softcover. **Price: $8.95**

TREE CARE MANUAL

The conscientious gardener's guide to healthy, beautiful trees. Covers planting, grafting, fertilizing, pruning, and spraying. Tells how to cope with insects, plant diseases, and environmental damage. 224 pages; 8½×11; softcover. **Price: $8.95**

UPHOSTERING

Upholstering is explained for the average householder and apprentice upholsterer. From repairing and regluing of the bare frame, to the final sewing or tacking, for antiques and most modern pieces, this book covers it all. 400 pages; 5½×8¼; hardbound. **Price: $12.95**

WOOD FURNITURE: Finishing, Refinishing, Repairing

Presents the fundamentals of furniture repair for both veneer and solid wood. Gives complete instructions on refinishing procedures, which includes stripping the old finish, sanding, selecting the finish and using wood fillers. 352 pages; 5½×8¼; hardbound. **Price: $9.95**

ELECTRICITY/ELECTRONICS

ELECTRICAL LIBRARY

If you are a student of electricity or a practicing electrician, here is a very important and helpful library you should consider owning. You can learn the basics of electricity, study electric motors and wiring diagrams, learn how to interpret the NEC, and prepare for the electrician's examination by using these books.

Use the order coupon on the back of this book.
All prices are subject to change without notice.

Electric Motors, 3rd Edition. 528 pages; 5½×8¼; hardbound. **Price: $12.95**

Guide to the 1981 National Electrical Code. 608 pages; 5½×8¼; hardbound. **Price: $13.95**

House Wiring, 5th Edition. 256 pages; 5½×8¼; hardbound. **Price: $9.95**

Practical Electricity, 3rd Edition. 496 pages; 5½×8¼; hardbound. **Price: $13.95**

Questions and Answers for Electricians Examinations, 7th Edition. 288 pages; 5½×8¼; hardbound. **Price: $9.95**

ELECTRICAL COURSE FOR APPRENTICES AND JOURNEYMEN
A study course for apprentice or journeymen electricians. Covers electrical theory and its applications. 448 pages; 5½×8¼; hardbound. **Price: $11.95**

RADIOMANS GUIDE, 4th Edition
Contains the latest information on radio and electronics from the basics through transistors. 480 pages; 5½×8¼; hardbound. **Price: $11.95**

TELEVISION SERVICE MANUAL, 4th Edition
Provides the practical information necessary for accurate diagnosis and repair of both black-and-white and color television receivers. 512 pages; 5½×8¼; hardbound. **Price: $11.95**

ENGINEERS/MECHANICS/ MACHINISTS

MACHINISTS LIBRARY
Covers the modern machine-shop practice. Tells how to set up and operate lathes, screw and milling machines, shapers, drill presses and all other machine tools. A complete reference library. **Price: $29.95**

Vol. 1—Basic Machine Shop. 352 pages; 5½×8¼; hardbound. **Price: $10.95**

Vol. 2—Machine Shop. 480 pages; 5½×8¼; hardbound. **Price: $10.95**

Vol. 3—Toolmakers Handy Book. 400 pages; 5½×8¼; hardbound. **Price: $10.95**

MECHANICAL TRADES POCKET MANUAL
Provides practical reference material for mechanical tradesmen. This handbook covers methods, tools equipment, procedures, and much more. 256 pages; 4×6; softcover. **Price: $8.95**

MILLWRIGHTS AND MECHANICS GUIDE
Practical information on plant installation, operation, and maintenance for millwrights, mechanics, maintenance men, erectors, riggers, foremen, inspectors, and superintendents. 960 pages; 5½×8¼; hardbound. **Price: $19.95**

POWER PLANT ENGINEERS GUIDE
The complete steam or diesel power-plant engineer's library. 816 pages; 5½×8¼; hardbound. **Price: $16.95**

QUESTIONS AND ANSWERS FOR ENGINEERS AND FIREMANS EXAMINATIONS, 3rd Edition
Presents both legitimate and "catch" questions with answers that may appear on examinations for engineers and firemans licenses for stationary, marine, and combustion engines. 496 pages; 5½×8¼; hardbound. **Price: $10.95**

WELDERS GUIDE
This new edition is a practical and concise manual on the theory, practical operation and maintenance of all welding machines. Fully covers both electric and oxy-gas welding. 928 pages; 5½×8¼; hardbound. **Price: $19.95**

Use the order coupon on the back of this book.
All prices are subject to change without notice.

WELDER/FITTERS GUIDE

Provides basic training and instruction for those wishing to become welder/fitters. Step-by-step learning sequences are presented from learning about basic tools and aids used in weldment assembly, through simple work practices, to actual fabrication of weldments. 160 pages. 8½×11; softcover. **Price: $7.95**

FLUID POWER

PNEUMATICS AND HYDRAULICS, 3rd Edition

Fully discusses installation, operation and maintenance of both HYDRAULIC AND PNEUMATIC (air) devices. 496 pages; 5½×8¼; hardbound. **Price: $10.95**

PUMPS, 3rd Edition

A detailed book on all types of pumps from the old-fashioned kitchen variety to the most modern types. Covers construction, application, installation, and troubleshooting. 480 pages; 5½×8¼; hardbound. **Price: $10.95**

HYDRAULICS FOR OFF-THE-ROAD EQUIPMENT

Everything you need to know from basic hydraulics to troubleshooting hydraulic systems on off-the-road equipment. Heavy-equipment operators, farmers, fork-lift owners and operators, mechanics—all need this practical, fully illustrated manual. 272 pages; 5½×8¼; hardbound. **Price: $8.95**

HOBBY

COMPLETE COURSE IN STAINED GLASS

Written by an outstanding artist in the field of stained glass, this book is dedicated to all who love the beauty of the art. Ten complete lessons describe the required materials, how to obtain them, and explicit directions for making several stained glass projects. 80 pages; 8½×11; softbound. **Price: $6.95**

Use the order coupon on the back of this book.
All prices are subject to change without notice.

BUILD YOUR OWN AUDEL
DO-IT-YOURSELF LIBRARY AT HOME!

Use the handy order coupon today to gain the valuable information you need in all the areas that once required a repairman. Save money and have fun while you learn to service your own air conditioner, automobile, and plumbing. Do your own professional carpentry, masonry, and wood furniture refinishing and repair. Build your own security systems. Find out how to repair your TV or Hi-Fi. Learn landscaping, upholstery, electronics and much, much more.

HERE'S HOW TO ORDER

1. Enter the correct title(s) and author(s) of the book(s) you want in the space(s) provided.

2. Print your name, address, city, state and zip code clearly.

3. Detach the order coupon below and mail today to:

Theodore Audel & Company
4300 West 62nd Street
Indianapolis, Indiana 46206
ATTENTION: ORDER DEPT.

All prices are subject to change without notice.

- -

ORDER COUPON

Please rush the following books(s).

Title _____

Author _____

Title _____

Author _____

NAME _____

ADDRESS _____

CITY _____ STATE _____ ZIP _____

☐ Payment enclosed _____
  (No shipping and   Total
  handling charge)

☐ Bill me (shipping and handling charge will be added)

Add local sales tax whe̶r̶e̶ 323 5/98   47

Litho in U.S.A.   31918